Praise for Defining the Wind

"[An] enchanting stroll through maritime and science history . . . [Huler is] a charming guide."

—*The New York Times Book Review*

"Huler writes with self-deprecating wit . . . he captures the Beaufort scale's 'open-hearted intellectual decency.'"

—*The New Yorker*

"Inspired by the invisible force around them, sailors have always tried to describe the wind. . . . As Hurricane Charley has just taught us, landlubbers, too, need to keep a weather eye. Hence the timeliness of *Defining the Wind,* Scott Huler's meteorological foray. Its principal figure is Sir Francis Beaufort (1774–1857), the son of a poor Irish parson who led a life out of a Patrick O'Brian novel: He went to sea young, fought the French, completed a survey of the southern Turkish coast and, after being wounded, retired to dry land, eventually heading the British Admiralty Hydrographic Office. . . . A roundabout voyage, but worth it."

—*The Wall Street Journal*

"[A] reminder of why we read science books. . . . Read Huler and you'll pay more attention to the air moving through your backyard, fluttering leaves, rattling windows. . . . Consummate examples of how a writer with enough determination can mine a deep vein of curiosity and use it to produce a compelling, powerful, and, yes, interesting book."

—*Boston Globe*

"On a scale of 0 to 13, this will blow you away . . . Hang on for a lighthearted romp over two hemispheres, two centuries of discovery, and the consuming passions of two men: Beaufort and Huler."

—*Seattle Times*

"[Huler's] thoroughly researched account of a marvelous collision of forces (natural, political, and creative) is as invigorating as a cool ocean breeze."

—*Entertainment Weekly*

"[Huler] comes off as a likeable captain, delivering punchy, humorous prose . . . while heeding his own words—that the human body is the 'greatest perceptive instrument ever designed'—by adding his fresh perspective to both history and lore."

—*Village Voice*

"To some people, the devil is in the details. But a true researcher will revel in the details. That's what Mr. Huler joyfully discovered in his investigations, and it shows in every page of this unusual and delightful book."

—*Dallas Morning News*

"An entertaining, informative, sometimes quirky read about the nature of wind and how for four centuries observers have tried to find a uniform way to describe it . . . a fascinating blend of history, biography, nautical lore, science, and personal journal."

—*Charlotte Observer*

"Parts history, mystery, and memoir, this take on ancient mariners measuring wind velocity could be the latest unlikely hit."

—*Southwest Spirit Magazine*

"[A] book about a 12-point scale used to measure the intensity of wind has changed my life. . . . The writing is clean and swift, the tone conversationally delightful. . . . [Huler] has done a magnificent job. Finally, this is a book of practical philosophy, about living your life alertly, being awake to the marvelous world around you. It's also in an artfully oblique way a handbook for every writer, amateur or professional, a refreshing primer on the most basic ethic of the craft."

—*Raleigh News & Observer*

"At [the book's] heart is a fascination with the language we use to describe the world around us. Less a piece of science writing than a writer's meditation on science, this gem of a book is equal parts history, mystery, and memoir . . . and deserves a wide audience."

—*Publishers Weekly* (starred review)

"[N]ot just a beautifully written portrait of Sir Francis Beaufort and of the scale that bears his name . . . it is also . . . a philosophical and historical tale of science as a way of observing and making sense of the world. Highly recommended."

—*Library Journal*

"Whether tracing the scale's evolving linguistic content or the route of one of Beaufort's surveys, Huler wonderfully relays the history contained, as he so aptly writes, in the Beaufort scale's 'one hundred ten words . . . and four centuries of backstory.'"

—*Booklist*

DEFINING THE WIND

The Beaufort Scale, and How a 19th-Century Admiral Turned Science into Poetry

Scott Huler

THREE RIVERS PRESS • NEW YORK

Published in the United States by Three Rivers Press, an imprint of the Crown Publishing Group, a division of Random House, Inc., New York.

www.crownpublishing.com

Three Rivers Press and the Tugboat design are registered trademarks of Random House, Inc.

Originally published in slightly different form in hardcover in the United States by Crown Publishers, an imprint of the Crown Publishing Group, a division of Random House, Inc., New York, in 2004.

Library of Congress Cataloging-in-Publication Data

Huler, Scott.

Defining the wind : the Beaufort scale, and how a nineteenth-century admiral turned science into poetry / Scott Huler.

Includes index.

1. Beaufort scale. 2. Beaufort, Francis, Sir, 1774–1857. I. Title.

QC933.H85 2004

551.51'8—dc22 2003025973

ISBN 1-4000-4885-0

Printed in the United States of America

DESIGN BY LEONARD W. HENDERSON

10 9 8 7 6 5 4 3 2 1

First Paperback Edition

To Lori

Contents

Nothing I am sure can be more useful than comparing our present ideas with those of old time, tracing back our chains of actions to their primary sources or motions, ascertaining the causes of our successes or failures, in short studying the history of our own mind.

—Francis Beaufort, Journal, 1805

Remember to get the weather in your god damned book—weather is very important.

—Ernest Hemingway, letter to John Dos Passos

beat ● because

BEAUFORT SCALE

BEAUFORT NUMBER	NAME	MILES PER HOUR	DESCRIPTION
0	calm	less than 1	calm; smoke rises vertically
1	light air	1–3	direction of wind shown by smoke but not by wind vanes
2	light breeze	4–7	wind felt on face; leaves rustle; ordinary vane moved by wind
3	gentle breeze	8–12	leaves and small twigs in constant motion; wind extends light flag
4	moderate breeze	13–18	raises dust and loose paper; small branches are moved
5	fresh breeze	19–24	small trees in leaf begin to sway; crested wavelets form on inland waters
6	strong breeze	25–31	large branches in motion; telegraph wires whistle; umbrellas used with difficulty
7	moderate gale (*or* near gale)	32–38	whole trees in motion; inconvenience in walking against wind
8	fresh gale (*or* gale)	39–46	breaks twigs off trees; generally impedes progress
9	strong gale	47–54	slight structural damage occurs; chimney pots and slates removed
10	whole gale (*or* storm)	55–63	trees uprooted; considerable structural damage occurs
11	storm (*or* violent storm)	64–72	very rarely experienced; accompanied by widespread damage
12	hurricane*	73–136	devastation occurs

*The U.S. uses 74 statute mph as the speed criterion for hurricane.

syn BEAUFUL, LOVELY, HANDSOME, PRETTY, COMELY, FAIR ***shared meaning element*** **: pleasing to the mind, spirit, or senses** ***ant*** **ugly**

The 110 best words ever written, in their natural habitat:
Webster's New Collegiate Dictionary, Ninth Edition.

DEFINING THE WIND

INTRODUCTION

September 6, 1996:

Hurricane Fran and Before

IN RALEIGH, 120 miles inland from the North Carolina coast, we mostly experience hurricanes on TV.

That's not to say a hurricane isn't a real event in central North Carolina. The day before a storm is due at the coast we often see the gently curving tendrils of feeder clouds at sunset, illuminated pink and yellow, giant across the blue sky looking exactly like they do on the weather radar screens—like galaxy arms, like propellers, like streamers. By the time we see those clouds we have been hearing about the storm for at least a week, in breathless weather-guy staccato, in clipped news-desk voices, in vague newspaper reports filled with horrid reminders of what happened the last time.

Then eventually the storm shows, and it hits South Carolina or moves up the coast or spends itself, whirling offshore for a few days, then drifts north and dies. If the storm itself does reach Raleigh, what we usually get is a very rainy day—a couple inches of rain and enough wind to knock down a tree branch in every other yard. With some storms the rain is truly horrendous and damaging, but even so, it's worse to the east, toward the ocean. Mostly our hurricanes come at us through shouted television reports from the weatherpeople hanging for dear life on to lampposts on the evening news, traffic lights swinging wildly above them to demonstrate that no kidding, it's real windy.

Still, a few years ago, in early September, Hurricane Fran headed north toward the Chesapeake, thought better of it, took a sharp left, and plowed directly up the Cape Fear River basin, suckerpunching Raleigh with winds of up to 79 miles an hour. That is, 120 miles inland, the storm was still by the textbook wind-speed definition a hurricane, and plenty of it.

That night found me out in my pickup truck, reporting on that storm for the local newspaper, so I can tell you that at two-thirty in the morning of September 6, 1996, Hurricane Fran left no room for doubt. After more than eight hours of dodging falling trees and twisted, sparking electrical wires, after hours of standing with police officers as the syncopated flashing blue lights illuminated the hysterically whipping trees like strobes, I was almost done for the night. The squad car in which I had been riding had been called in as the police gave up in the worst of the storm, and as the wind continued to rise I had picked my way through the maze of obstructed highways and flooded dips. Near my house I grew overconfident and steered down a nearby hill for one last look at the madness, and soon found myself sitting in my pickup not five blocks from home. I couldn't drive forward—at a Y in the road, my headlights garishly illuminated collapsed trees that blocked both forks. I would have to back the truck, turn around, and slink out below leaning utility poles and thrashing trees. Before I backed away, though, I sat a moment to watch.

Raindrops lashed through the headlight beams horizontally, illuminated like tracer bullets; the wind flung whole emerald trees back and forth like a child shaking dandelions to blow away the down. Entire boughs shook free as easily as those dandelion seed puffs, and as I looked out my windshield I saw fragments of Raleigh flying sideways through the zinging streaks of rain. Here a 12- or 16-foot limb of a tree, there an aluminum piece of storefront, then a road sign or lawn chair lifted like a kite over two or three front yards. Utility transformers exploded every few moments—a blue flash of light and a low boom. The sound of breaking limbs cracked like sniper fire. People always say that in a tornado the wind sounds like a freight train going by, and that turns out to be true. We have freight trains going through Raleigh, and when they're going good and fast, they have a particular throaty roar.

During Hurricane Fran we heard that sound for seven hours straight. As I sat, the wind would get underneath my pickup and I would feel the truck lift up on its wheels and shimmy, like a linebacker doing a footwork drill. It was a terror to be out in it, and something like a privilege—like seeing the Grand Canyon or the Himalayas.

It was also dangerous, so I backed my truck up—even the police had called in their guys, and as the worst of it passed, everybody was just taking cover and waiting until morning. I crawled the few blocks home, discovered to my delight that my pecan tree was still standing, and climbed the stairs to go to sleep. The overwrought trees delivered a relentless beating to the roof, though, which sent me downstairs, where I fell asleep during the spooky calm as the eye of the storm passed directly overhead. I awakened briefly sometime later to renewed wind and the noise of something crashing into the roof and rolling down the steep pitch—I heard its heavy landing, whatever it was, in the soaked leaves at the side of the house. Electricity was long gone—a nervous foray upstairs with a candle revealed no holes in the roof, and I went back down to fitful sleep.

In the morning, like other wide-eyed Raleighans I crept out of my door and surveyed the devastation. No street was passable; trees crossed every road, rendering driving almost unimaginable. Roofs had flown from houses, doors had ripped loose from their moorings; furniture and detritus filled yard and street like tumbleweed. Still, I had been lucky. Everywhere, downed trees had crushed cars, shattered doors, pierced roofs. Yet the noise that awakened me had been the wind ripping only the steel bonnet off my chimney and sending it toppling down the sloping roof of my house; I found it out behind the house, among boughs and the flung debris from a half-dozen neighborhood yards. I stood for a long moment, looking at the twisted bonnet lying in the pine straw and soggy leaves. Something—it reminded me of something. I wandered over to a neighbor boiling water for coffee on his outdoor grill, but it nagged at me.

Not long afterwards I remembered, and I ran home, and to the bookcase, for my dictionary. I paged quickly to the Beaufort Scale, and I knew then at least how fast the wind had been raging in the middle of the night when it upset my chimney.

> Beaufort Number 9—Strong gale: 47–54 miles per hour. Slight structural damage occurs; chimney pots and slates removed.

That's what it was, at least. A metal bonnet is as close as I'm going to come to a chimney pot (which according to the dictionary is usually a cylindrical masonry pipe, though it can be made of metal). But in any case the wind had been strong enough to curl underneath it, rip the twisting masonry nails right out of my brick chimney, and send it tumbling. It was a strong gale—a little faster than a fresh gale (or just plain gale), which goes 39 to 46 miles an hour and "breaks twigs off trees; generally impedes progress." A little weaker by that point, perhaps, than a whole gale (or storm), which goes up to 63 miles an hour and describes what the entire city of Raleigh looked like that stunning morning: "Trees uprooted; considerable structural damage occurs."

I was applying the Beaufort Scale.

I had waited almost fifteen years to do it.

TO EXPLAIN WHY I came to love the Beaufort Scale, I have to explain how I came to know it in the first place, and that starts not in a hurricane or a fresh gale or even outdoors, but in a small office near the Cooper River in Pennsauken, New Jersey. For many years I was a copy editor. That's good honest work and underappreciated, but above all it's a great place to learn how writing works. From character—is this the right punctuation mark? is this word spelled correctly?—to clause, from sentence to paragraph, from passage to complete manuscript, a copy editor tinkers with prose like a jeweler with a watch. It's great experience, and great training for a writer. Learning to copyedit before becoming a writer is like being a mechanic before learning to drive a race car. The understanding of the secret processes behind the magic can only help, especially when the handling gets rough.

This was around 1983, and I was copyediting at a small technical publisher near Philadelphia, where I lived. We elbowed our way through prose describing these new machines called minicomputers. We described them for businesspeople—few consumers were buying them yet. We used actual blue pencils to edit actual typewritten words on actual pieces of paper—it was a long time ago—and our little warren of four editors was considered something of an ivory tower. Certainly we ensured that abbreviations were used properly and units of measure applied correctly, but that was only the smaller part of our job. We spent much more of our time worrying about the distinction between *that* and *which,* about "not only . . . but" clauses, about the proper use of a semicolon. About the difference between *ensure* and *insure* and *make sure.* We used traditional editing and proofreading marks, taking a weird satisfaction in mastering those arcane symbols.

We excised excess, declared war on danglers, and devoured usage manuals, rereading *The Elements of Style* and *The Chicago Manual of Style* and *Words into Type* repeatedly and gladly. We each had a personal copy of the same dictionary—the Merriam-Webster *New Collegiate Dictionary,* Ninth Edition—standard in editorial departments nationwide.

It doesn't sound like much—your own dictionary. But when your job is riding herd on the spelling and usage of others, a dictionary becomes far more than a tool. It's a kind of helper, a resource of first and last resort. If a copy editor considers him- or herself a word wrangler, a knight errant—and almost all of them do—the dictionary is the squire, the doughty sidekick providing invaluable support.

Of course, we did more than merely look up words in our dictionaries. We pored over them, annotated them, left blue-pencil jokes in one another's copies (the picture of a platypus in mine sported a lovely blue-pencil plaid coat—a *plaid*ypus—compliments of a highly entertaining copy chief).

Above all, we read them.

Not just because words are lovely and copy editors love them, and not just because one could commonly profit from the words onto which one thereupon stumbled, but because the dictionary represents a kind of breathing example of the copy editor's craft: It makes its living by conveying meaning, in as few words and as simply as possible. To a technical copy editor, a dictionary is the Holy Grail.

Copy editors like to recite almost from memory the passage in *The Elements of Style* that says "a sentence should contain no unnecessary words, a paragraph no unnecessary sentences, for the same reason that a drawing should have no unnecessary lines and a machine no unnecessary parts." It defines the goal, in a copy editor's eyes, of all good writing: vigor, conciseness, clarity. To a copy editor, especially a technical copy editor, the best writing approaches the vanishing point. The fewest words to get the idea across—to get the idea across *exactly.* If you could reduce writing to a sort of elementary symbolic logic, copy editors would love that and would find it beautiful.

For that reason, I think, copy editors love lists and symbols, and the dictionary is full of those. Things like the lists of words following *anti-* or *pre-* thrill copy editors because they are so simple and so concise: no unnecessary definitions, just a list. Gets the point across.

And look through the Merriam-Webster Ninth *New Collegiate Dictionary* and you will find countless such lists, examples of pure, informative communication in the miraculous shorthand of a list, an order, a scale—the alphabets of the world; the languages of the world; the calendars of the world. Lists of poker hands, orders of classical column, types of money, parts of a fish. If you look up *clouds* in the dictionary, you will find a simple, tiny drawing of the ten or so different types of clouds, and that's all you need to know—concise and nearly perfect. Musical scales, Morse code, the elements—periodically or alphabetically, depending on where you look—all find their purest and simplest descriptive form in the dictionary.

So, it turns out, does the wind. The wind is described to perfection—reduced to its essence like distillate in an Erlenmeyer flask—in the Beaufort Scale. The Beaufort Scale is the quintessence of that verbal economy, the ultimate expression of concise, clear, and absolutely powerful writing, 110 words in 6-point type. In fact, the Beaufort Scale description of the wind doesn't merely reach that highest perfectible level of clarity. As may be necessary, reaching that level, it surpasses it and becomes poetry.

I ran across the Beaufort Scale for the first time one day in that editorial office while looking something up. I don't remember what I was trying to find, but I know I didn't find it, and I know some report on an early computer workstation lay unedited on my desk the remainder of the day. The Beaufort Scale entirely captivated me.

It's given in a little table: Beaufort numbers from 0 to 12 down the left, followed by names and velocities of each wind category, and finally, along the right, a brief description of each. Less than one mile per hour is Beaufort 0, "calm," with the description, "smoke rises vertically." By force 1, "light breeze," with a velocity of one to three miles per hour, "direction of wind [is] shown by smoke but not by wind vanes."

Count those—that's fifteen words, and I already had a complete visual image. I could see the little tiny village, in imagination probably in a green New England valley. There is a little house, with a tiny wisp of smoke coming out of the chimney, curling up at the slightest angle, as though drawn by a child. Nearby is a church, with a steeple, atop which is a windvane. By Beaufort 2, "light breeze," the air moves between four and seven miles per hour and the description says "wind felt on face; leaves rustle; ordinary vane moved by the wind." Another twelve words—twenty-six in total—and we've added trees on the town green, plus a barn in the distance, atop which is an ordinary windvane, probably in the shape of a rooster. As readers we're on that green, feeling the wind on our faces. In fact, in those twenty-six words we've engaged four of our five senses—the sight of the windvane's motion, the sound of the leaves rustling, the feeling of the wind on our faces, and, presumably, the nice autumnal smell of the curlicue of smoke wafting from that chimney.

I had never read anything quite like this. I was awestruck by its economy, by the vigor of its prose, so I showed it to the other editors on the staff. They politely smiled and nodded and returned to their editing, but not me. I felt I had discovered a species of gem, a small miracle, and I pored over it like a psalm or a koan.

I eventually did get back to work, but I soon returned to the Beaufort Scale, and I returned often. Before long I fell into the habit of reading it regularly, returning to it as to a treasured poem or a favorite passage in Aeschylus or the Bible. I sought it especially when editing notably turgid copy, filled with the jargon and cant and vagueness that characterize writing about the new computer technology in which the company for which I worked specialized. Somehow the Beaufort Scale began to represent for me a kind of anti-technobabble, the exact opposite of the flabby, empty prose I was paid to edit. I became something of a pest on the subject, peddling the Beaufort Scale to other editors and writers. I sang its praises and counted its words, and by the time I left that job—I had it only eighteen months or so—I could be depended on to claim that the Beaufort Scale as I found it in that Merriam-Webster Ninth *New Collegiate Dictionary* was the best, clearest, and most vigorous piece of descriptive writing I had ever seen. As far as I was concerned, the Beaufort Scale—it actually turns out to be the original version of the Beaufort Scale Land Specification, written in 1906, almost exactly a century after the scale was first devised, but I could hardly have known that back in 1984—represented, in thirteen entries comprising 110 total words, the apex of descriptive nonfiction in English.

As I said, I didn't keep that job as a copy editor all that long. But in my next job, and my next, and every job since, I've returned to the Beaufort Scale time after time, and it became something of a hobby. After I returned to the Beaufort Scale often enough, I became interested in its author, and there, for the first time, the Ninth *New Collegiate* let me down. The definition of the Beaufort Scale contains the bracketed aside "[Sir Francis *Beaufort*]," but nothing more. The index of biographical names that appears at the end of the dictionary—between the list of foreign words and phrases and the lists of geographical names—told me only that Sir Francis Beaufort was a British admiral who had lived between 1774 and 1857.

That wasn't enough. I wanted to know more about this stylist, this genius of descriptive prose and the scale he created. I wanted to know where the Beaufort Scale came from.

As it turned out, my pursuit of the origin and history of this odd little piece of descriptive prose would lead me in some pretty surprising directions. On the trail of Beaufort and his scale I would run into a parade of unexpected people—from Daniel Defoe to Charles Darwin; from Anders Celsius to John Smith (the Pocahontas guy); from the Elgin Marbles to Seamus Heaney; from Tycho Brahe to Captains Bligh, of the *Bounty,* and Cook, of the *Endeavour.* I would learn about the British Admiralty and about one of the greatest machines of all time, the sailing ship; about the vast undertaking of mapping the coastlines of the world; I would learn about the force of the wind, and about early attempts to understand how our planet works; I would learn about writing, well and otherwise. I would learn about the essential human undertaking of categorization. But above all, I would come to consider the nature—and the necessity—of observation, both among the gentleman scientists of the nineteenth century and among the Weather Channel viewers of today. As I pursued this little gem of writing, I would eventually find that what drew me to it may have been, above all, the kind of life it implied its writer led.

But that was all to come. After Hurricane Fran sent me to my dictionary, I knew only something very simple.

I wanted to know Sir Francis Beaufort.

CHAPTER 1

Beaufort of the Admiralty

THE WORST THING about the ferry that runs between Buenos Aires, Argentina, and Montevideo, Uruguay, is that it's a hydrofoil. On the one hand this makes for a fast trip—kicking up twin spumes of seawater, the hydrofoil takes only two and a half hours to go 137 miles across the Rio de la Plata, the world's widest river. But on the other hand, the speed of the boat means they won't let you outdoors. A boat going fifty-five miles per hour kicks up quite a wind (it's Beaufort force 10, if you're wondering: "trees uprooted; considerable structural damage occurs" if you're on land, to say nothing of the terror it would inflict on a straw hat and sunglasses), so the speed prevents you from doing what you naturally want to do on a boat in summer, which is stand on the deck and watch the world go by.

Or, in my case, watch the world come toward you. What I specifically wanted to see was Montevideo, and I wanted to watch it emerge from the vastness of the Rio de la Plata and assert itself on my eye as we approached it from the west, exactly as Sir Francis Beaufort had in 1807. It shouldn't have been hard to see what he wanted me to see—he actually left me directions, and I boarded the boat with a sheaf of maps under my arm and high hopes.

But no soap. The ferry *Juan Patricio* doesn't even have a deck open to the weather, so the tourist-class passengers are left squinting out of windows with an aerodynamic slant seemingly designed to maximize the reflection of interior seats, carpeting, and the knees and feet of

Francis Beaufort, a youthful fifty-four in the year the scale was officially adopted by the Admiralty. Black and red chalk, drawn in 1838 by William Brockedon.

By Courtesy of the National Portrait Gallery, London.

passengers. The window ledge itself reflected a sharp line at almost exactly horizon height as you gazed out the window, and trying to keep the two lines separate, in a boat skipping over water at 55 mph, can actually leave one rather queasy.

I initially envied the first-class passengers, whom I imagined strolling in front of huge banks of flat windows as the broad Rio de la Plata sped regally by beneath the boat, but an earnest look at the attendant, a flourished sketchbook, and a few words of stammered Spanish got me, an accomplice, and my maps upstairs, where I found the passengers had a better snack bar and much wider seats—but no view out the front of the boat at all. So it was back down to steerage and more attempts to find an angle at which I could peer through the windows and see something other than my own annoyed expression reflected back at me.

I ENDED UP TRAVERSING the Rio de la Plata solely because Sir Francis Beaufort had been there before me—he was there in 1807, I in 2002—and I yearned to go where he had been, to see what he had seen.

This wasn't what I had expected. From the scale that bears his name I had naturally presumed Sir Francis—Queen Victoria made him a Knight Commander of the Order of the Bath in 1848, hence the "Sir"—was either a writer or a meteorologist, so I was surprised to find that the standard biography called him *Beaufort of the Admiralty.* My girlfriend loved the stentorian sound of the title; if she saw me lying on the couch reading it she would wave her arms magisterially and intone "Beaufort . . . of . . . *the Admiralty,*" and I would find it hard to maintain my sense of purpose. The title, though, is exactly right: It's the type of book that instead of merely introducing the first captain under whom Beaufort served—his name was Lestock Wilson—also includes information surrounding Wilson's birth. It's the type of book that has endnotes to its footnotes. If you look in its bibliography, you'll find books like *Mr. Barrow of the Admiralty,* so the title really just keeps up tradition. In some ways, of course, I had been right about Beaufort: he *was* interested in the weather, and he wrote incessantly, if not especially well, filling letters, logs, and journals—and even a book—with prose that swung from the pole of engineer-flat to the tropics of almost absurdly purple without spending much time in the temperate zones.

But above all, Beaufort was a sea captain for the Admiralty—the British Royal Navy—at a moment that put him at the center of one of the greatest cartographic enterprises the world has ever known. In the nineteenth century, while the British merchant and naval fleets were busy taking over the world, Sir Francis Beaufort was the man responsible for making sure they knew what they were getting—for creating charts of the coastlines of the world. The Hydrographic Office of the Admiralty was creating the Admiralty Chart.

An Admiralty Chart, of course, is a map of the sea, or a particular portion of the sea. If you're sailing to, say, Montevideo, you'll get yourself charts prepared by some trustworthy organization—the British Admiralty, the Argentine or United States government—and you'll start planning. The charts showing the largest areas—say, the South Atlantic Ocean—have scales up to about 1:5,000,000, at which size you can see land on both sides of the narrowest part of the Atlantic. By the time you're heading into port at Montevideo, you'll be looking at scales of 1:10,000 or even smaller—close enough to give you soundings demonstrating the best approach to the channel into the harbor, and even showing the docks in easy detail.

An Admiralty Chart, however, is a good way to find your way from Buenos Aires to Montevideo in exactly the same way that the *Oxford English Dictionary* is a good place to go to find out whether *desperate* is spelled with an *e* or an *a:* it can do that, but it can do an awful lot more. The Admiralty Chart is the gold standard, the true north, the understanding of centuries of navigational and hydrographical lore, distilled into two dimensions and six square feet. "Put your faith in God and your trust in the Admiralty Chart," nineteenth-century sailors would say, and it's worth noting that at that point the Admiralty Chart probably had better long-term prospects. Hydrography is the seagoing sibling of cartography—the cartographer maps the land, the hydrographer draws the coastlines and sounds the sea. Beaufort spent the second half of his career as hydrographer to the Admiralty, the period regarded as the high noon of hydrography. When those sailors resolved to place their trust, they were trusting Sir Francis Beaufort.

ONE WAY TO JUDGE THE MAGNITUDE of Beaufort's contribution to the Admiralty Chart is to compare it with the Beaufort Scale. I found the Beaufort Scale in the Merriam-Webster dictionary, but that's just where I happened to stumble across it. You can find it anywhere.

Open any book about the weather—any book—and you'll find the Beaufort Scale. Look in any navigation manual—there it is. The United States Navy teaches the Beaufort Scale to its midshipmen and virtually every reasonable sailing course includes training in its use.

Kite companies include the Beaufort Scale in their advertising literature. Websites for weather organizations and television stations include the Beaufort Scale.

The *Farmer's Almanac* contains the Beaufort Scale.

Which makes it somewhat surprising that Beaufort's obituary neglected to mention it even once. And when I say obituary, don't think of the obituaries you're used to seeing in modern newspapers, with a few lines about community organizations and surviving children. On December 17, 1857, in London, Sir Francis Beaufort—admiral, member of the Royal Society, Knight Commander of the Bath, member of an almost staggering number of other scientific societies—died at age eighty-four, surrounded by his family. Less than a month later, the London *Daily News* ran an obituary, even an appreciation, of Beaufort that received sufficient interest to be reprinted as a pamphlet.

In its reprint version, the encomium runs to sixteen pages.

It never mentions the Beaufort Scale of wind force.

Not once. It discusses the education Beaufort's father provided him. It describes a book he wrote in 1818, it includes a venture in telegraphy in which Beaufort played a role (though it neglects to point out the total, embarrassing failure of that venture), it tells about the first time Beaufort nearly died (which, it turns out, he did quite a lot). But it never mentions the Beaufort Scale.

What it talks about in detail—sixteen pages of detail—is Admiralty Charts. Francis Beaufort made the contribution for which his peers remembered him as a hydrographer. In some ways he was born to it.

FRANCIS BEAUFORT was born, the middle child of seven, to an Irish family in County Meath in 1774. His father was Daniel Augustus Beaufort, descended of fallen French nobility of the Holy Roman Empire. Daniel Augustus was a distracted country parson who dabbled in everything from farming to architecture, even working as a magistrate; church buildings he designed still dot the Irish landscape, and he had bachelor's and master's degrees from Trinity College in Dublin. He eventually received an honorary LL.D. The Beauforts' was the kind of household where French was spoken and the family went to the theater and art exhibits, to readings and demonstrations. Francis's brother grew up to become a classical scholar, his sister a fine artist. As a biography of Daniel Augustus Beaufort says of the family, "no collection of pictures or prints was overlooked, no new play unpatronized, no Handel commemoration unheard, no library, museum or great edifice unvisited." When the first Irish hot-air balloon flight occurred in 1785, the Beauforts were in the crowd—and not long afterwards, Daniel Augustus made and launched a model of the balloon. It caught fire. With an architect for a father, all the children were taught to draw, and when Francis demonstrated an interest in the sea, he was sent to learn the basics of navigation.

Interested in just about everything, Daniel Augustus Beaufort had one remarkable success: he undertook in 1792 to make the first useful complete map of Ireland—though calling it complete is something of an overstatement. Not that it left off natural landmarks like rivers or roads, lakes or harbors; it's just that as a Church of Ireland parson, Beaufort made a map that included every religious structure in Ireland related to his denomination—meanwhile never once including something as inconsequential as, say, Christ Church Cathedral in Dublin, built in 1240.

Still, it appears that the Irish could find their way to church without a map, so the lacunae did not affect its sales. *Memoir . . . of Ireland illustrating the Topography of that Kingdom and containing a short Account of its present state civil and ecclesiastical* sold thousands of copies and went into several editions.

In fact, young Francis Beaufort even made some of the observations of latitude and longitude for his father's map, after taking several months of study in astronomy with Dr. Henry Ussher, a friend of his father who taught at Trinity College, Dublin. He was identified in the *Memoir* as "a pupil of Dr. Ussher."

Francis had only occasional traditional schooling, largely because his family had sometimes to flee its home. If the map of Ireland was the first successful map made by the Beaufort family—and the first of many—it was regrettably about the only undertaking Daniel Augustus ever brought to financial success. In the years before he made his map, he had been forced to move his family several times, often overseas to England and Wales, to keep a step ahead of debt collectors.

Yet from the letters Francis and his father exchanged, it appears that the scattered early life Daniel Augustus Beaufort created for him had no ill effects on Francis. All Francis took from his father was the ability to respond to new situations, a vast curiosity—and a profound love of physical observation. He loved to look around—and to write down what he saw.

And from the memories of his family, it appears that Francis Beaufort wanted to make his physical observations from the deck of a ship, and that he wanted that for as long as anyone in his family could remember. "At the age of five," his sister Louisa wrote years later, he "had manifested the most decided preference for the sea, had even refused to learn Latin or any of the rudiments of a learned profession & uniformly persisted in choosing a Naval life"—though she noted that there was no reason to believe Francis at the age of five had ever even seen seawater. Just the same, his father provided for him that training in astronomy essential for navigation, and in 1789, at fourteen, Francis Beaufort set sail as a sort of officer-in-training aboard the *Vansittart*, an East India Company tradesman bound for China and the Indies.

At least on the front end, the *Vansittart* had a reasonably successful voyage. It rounded the Cape of Good Hope and made its way to Jakarta, then called Batavia. Chosen to help the captain make astronomical observations to fix the latitude of Batavia, Francis found the observatory there deplorable and wrote home to his father that "a person walking about on the stairs or in . . . any room of the house shakes the Horizon and makes the objects turn about in the Equatorial Telescopes like anything." A central goal of the *Vansittart*'s journey was to survey the Gaspar Strait, just off Sumatra between the islands of Bangka and Belitung; sister ships of the East India

Company had been lost there on dangerous and poorly charted shoals, which the *Vansittart* was to find and chart. Young Francis had been taking bearings and making calculations throughout the journey, and was a help as the ship made the Gaspar Strait survey.

The *Vansittart* found the shoals, all right—it found them good, running aground and sustaining enough damage that it took on water so rapidly that the crew had to abandon the ship on a reef off a tiny island in the Java Sea. The Malay waters were filled with pirates, so the crew, terrified and perilously short of water, determined to make for a Dutch settlement called Palembang on the northeast coast of Bangka. In the hopes of returning to reclaim the ship's treasure, they threw overboard thirteen treasure chests and piled into open boats, hoping for the best and heading for the Dutch settlement. Along the way one of the boats, with five aboard, became separated and was lost; a vicious squall whipped the remaining boats, and later a group of pirate ships threatened to attack. The crew made it to Palembang, though, where they were relieved to find two British ships anchored in the harbor.

Captain Wilson persuaded the captains of those ships to ferry his crew back to the wreck of the *Vansittart* to recover what they could, though by the time they reached what remained of the ship, it was just as they had feared: Malay pirates had burned and pillaged it. The crew managed to recover only three of the treasure chests. After that, in the manner of the day, crew members scattered among other ships to make their way back to England. Beaufort ended up in Canton, where he occupied himself for two months making astronomical observations before sailing home with Wilson, arriving back in England a little more than a year after he had set out.

The news of the *Vansittart*'s sinking preceded him, though, and worried his father. There were no news or wire services in 1790, of course. The *Vansittart* sank on August 24, 1789, and its surviving crew reached safety in present-day Indonesia a couple of days later. At around the same time, it turned out, another sailor, after misadventure at sea and his own epic journey by open boat, had himself reached safety on Java, only a few miles away from the *Vansittart*'s misfortune. The news of the fate of the *Vansittart* reached England through the offices of this other sea captain, who had suffered not shipwreck but mutiny.

His name was William Bligh, and he had been captain of the *Bounty.*

SO WAIT A MINUTE. I go looking for the story of the guy who wrote this awesome wind scale that blew my mind. I start reading about his life, and before he's sixteen years old I've already run across a family's flight from the poorhouse, an early balloon flight, an eccentric father, a young man at sea, Malay pirates, shipwreck, castaways, buried treasure, and Captain Bligh, fresh off the mutiny on the *Bounty.* Not a single word about the wind, but honestly, at this point, who cares?

And what makes Beaufort most remarkable is that what he took away from all this adventure, what most influenced him throughout the rest of his life, was his delight in taking bearings. Surrounded by fighting and pirates and treasure and trading and storms, Francis Beaufort's heart led him toward his sextant and his level, his pen and his ink—toward the tools of the hydrographer's trade. Francis Beaufort was a man who liked to know where he was.

"Everyone has a hobby or his insanity," he would eventually write in his journal in 1806. "Mine I believe is taking bearings for charts and plans." He wrote those words sixteen years after his first sea journey, aboard the HMS *Woolwich,* of which he was commander—the ship that eventually took him to Montevideo, where nearly 200 years later I would follow in his wake. But as I read the story of his first journey, it was already clear to me that the spirit of the wind scale lay in

Beaufort's love of making those navigational observations, in his contribution to the mapping of the coastlines of the world.

WHEN I LEARNED that a journalism fellowship would take me to Buenos Aires, across the river from Montevideo, I ordered a copy of the 1808 Admiralty Chart of Montevideo based on Beaufort's surveys and drawings; it arrived rolled up in a tube and tied with a piece of cloth, so even unrolling it was an act of moment. When I did, I was stunned: It's a work of art. If you're approaching Montevideo by sea, Beaufort's chart gives you everything you need at a glance. About three feet long and a foot high, the map is broken into four parts—at left, an overview of the harbor, the city, and the outlying shoals and islands; next, a more detailed plan of the mountain, city, and harbor; a close-up of the outlying Flores Islands; and, along the top, drawings in Beaufort's hand of "The view from the outer Anchorage . . . of the Harbour of Monte Video. The Mount and the Light house on the left, the Town and Citadel on the right."

I thought the map was like a visual Beaufort Scale—all the information I could need about arriving in Montevideo by sea, distilled and reduced, clarified to its essence. I used a reducing photocopier to make a smallish copy of the map, and then when I ended up on that enclosed ferry, when I wasn't squinting out the window and trying to pick up the outline of the coast as we steamed ahead, I gazed at the chart. At that point I scarcely needed to, though—I had pored over it so closely since planning my trip that I felt I would recognize the coast the moment I saw it, like going to the airport to meet a relative you've seen all your life in photographs.

Beaufort's is an early Admiralty Chart. The Admiralty Hydrographic Office, founded in 1795, had existed for little more than a decade when Beaufort drew his chart in August 1807, and there were still few rules for the organization of such a chart. Latitude and

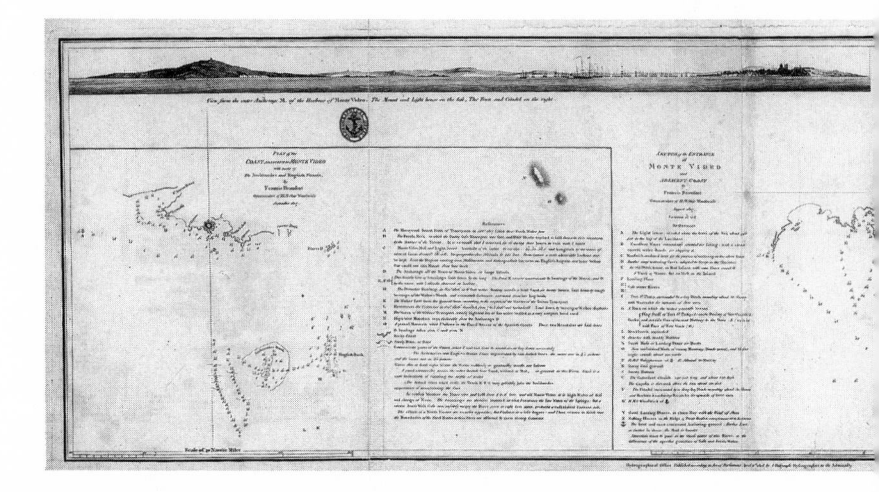

longitude are not given, for example, though Beaufort has included the height above sea level of the top of the mountain from which the city gets its name and of the cathedral church in the town, then still called a citadel.

Beaufort by then was working for the government. After one trip and one wreck, he decided he had had enough of merchant service, and joined the Royal Navy. He soon added to his shipwreck story a near drowning; participation in the British naval victory against the French fleet in what was termed the Glorious First of June, 1794; and a few saber wounds and a point-blank volley in the face from a blunderbuss while leading a gang of marines boarding a Spanish ship.

It was an exciting time. In fact, Beaufort's first brief time in charge of a ship was aboard *La Bonne Citoyennne,* a French corvette that his ship, the *Phaeton,* took in battle in 1796. ("A pretty thing she was," Beaufort wrote in his journal.) Renamed the *Speedy* by its new British owners, the ship served as the model for the sloop *Sophie* in *Master and Commander,* the first of Patrick O'Brian's famous series of sea

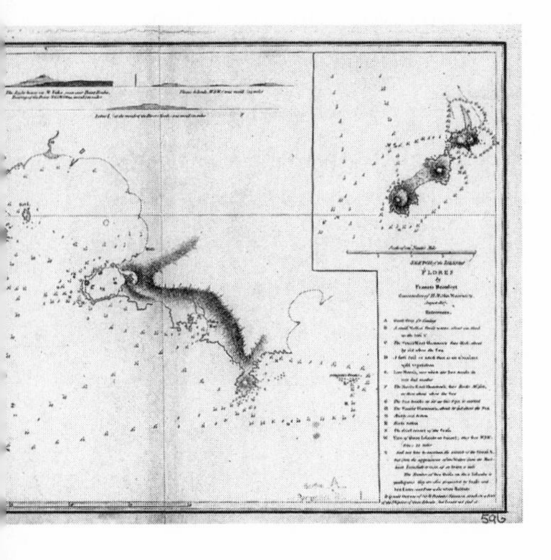

Beaufort's Admiralty Chart of Montevideo shows the Admiralty Chart just beginning to spread its wings.

novels. The *Speedy*'s subsequent commander, Thomas Cochrane, served as the model for O'Brian's Jack Aubrey. But in fact, though Beaufort was not the model for Aubrey, he almost could have been. A short, chipper man, Beaufort loved his military regalia and was a highly spirited but vastly competent man on the ship. He occasionally tweaked the authority of his commanders, once trying to avoid an onerous trip by leaving port with a fleet to which he knew he did not belong, only to be forced back into port by bad weather. Called on the carpet by his superiors, Beaufort cheerfully brazened it out on a technicality.

By 1806 Beaufort had reached the rank of commander of the *Woolwich,* a store ship dispatched to supply the British troops who had taken Buenos Aires from the Spanish that same year. By the time the *Woolwich* arrived, the British had soundly lost Buenos Aires right back, and had retreated across the Rio de la Plata to Montevideo, where, under the rules of engagement of the time, they were allowed to regroup as long as they promised to leave within a couple months.

Beaufort filled his time predictably, surveying Montevideo and the surrounding waters. "I seem to understand surveying harbours and shoals better than philosophizing," he had written to his father a few months before. "And what people do best they generally like best." With no other employment to distract him, Beaufort created a survey so thorough that another captain commented soon thereafter that Beaufort "did more in the month he was in the River Plate to acquire a correct knowledge of its Dangers, than was done by everyone together before." Though Beaufort had filled books and paper with surveys and bearings on every voyage of his life, this was the first of his surveys to be reproduced as a chart, and it made his reputation.

A glance at the chart shows why. The maps, surveys, and soundings Beaufort made in and near Montevideo were engraved virtually without change into the Admiralty Chart. The soundings are clear and accurate, the maps—bird's-eye views, with old-fashioned hachure lines, the predecessors to contour lines, showing the steepness of the mountain and hills—are crisp, and the legend simple and thorough: A wiggly line is rocky coast, a dotted one sandy bays, and specific shoreline landmarks are labeled: A, the lighthouse, or "K. *A* Rock *on which the* water *generally* breaks."

But most arresting by far is the sketch. Here Beaufort has drawn a panorama of Montevideo, just as he describes it: a mountain on the left, or west; to the east, a perfectly sheltered cove, into which juts, from the north, the town and citadel of Montevideo. You can see the lighthouse atop the mountain, ships in the harbor, with sails furled and pennants waving. On the right is the town, with the twin cupolas of the cathedral church visible in the beautifully engraved view.

But there's even more. Besides the plain hydrographical surveys that he made in his brief sojourn in Montevideo, Beaufort also undertook, for almost no reason other than that it was something he enjoyed, to create an "Eye-Sketch of Monte Video, Chiefly laid down by Pacing the Distances." That is, to fill his time, Beaufort went ashore, entered what was then the walled city, and literally paced the streets in order to make a map that, it seemed, nobody needed. He said he made it "to keep the Devil out of my mind."

To me, the harbor sketch and the map seemed a clue directly into Beaufort's mind, devil or no. The adventurous, ambitious young captain, stuck in port, looked around and found something worth doing: observing his surroundings and representing them so that others could understand them, too. This had nothing to do with the wind, but again, a map was a sort of visual equivalent of the wind scale, the most economical way of conveying the greatest amount of information. It's a beautiful thing, and especially beautiful for its plain utility. Beaufort wasn't creating fine art—he was drawing Montevideo so that captains after him could safely negotiate its shoals and find their way into its harbor, and his drawing accomplishes that goal. It brings to mind again Strunk and White's adjuration that "a sentence should contain no unnecessary words . . . for the same reason that a drawing should have no unnecessary lines." Beaufort's first Admiralty Chart has no unnecessary lines; it offers only economy, vigor, and clarity.

I believed it would lead the way to the clarity and economy I found in the Beaufort Scale, and I determined that by standing on the deck of a boat and seeing Montevideo as Beaufort had seen it, I might understand him, might come closer to his way of seeing. Inspired by his sketch of the harbor, I even took a drawing class, so I could learn the skills to take all that visual information and present it on paper, as Beaufort had. I envisioned myself pacing the deck as we slowly approached the harbor, eyes narrowed, sketching the coastline as he had done two hundred years before.

And then I found myself on that hermetically sealed hydrofoil with slanted windows and no deck. Seeing no options, I sat in my front-row seat, stared at the reflections, and sulked as I waited to see

Montevideo. In the lounge, a television screen displayed the nautical chart—in this case issued by the Argentine navy—of the Rio de la Plata as it went by, and it plotted the passage of the boat. With Beaufort's Admiralty Chart on my lap and a window separating me from the wind and sea, it made me almost despondent to realize my best view of the approach of Montevideo was going to be on a televised diagram.

Fortunately a friend in my traveling party wasn't so easily discouraged, and we engaged our hiccupy Spanish, talking to the ship's information officer. A sketchbook helped her get a general sense of my mission, but when I spread on her desk my photocopied 1808 Admiralty Chart and Beaufort's sketch of Montevideo, her eyes lit up. She picked up the phone, chatted, then nodded at me.

"*Sí,*" she said. "*Regresare en media hora.*"

A half hour later we were back. Another quick phone call, and she led us through first class to the door to the pilot house. She touched a few numbers on the button combination and we entered.

Captain Fernando Giunta greeted us—he had time for pleasantries, since a computer was mostly steering the boat—and when I pulled out my map, he brightened. He looked at the map, his navigator joining him to point to landmarks so familiar to them—here the mountain, the harbor, there a list of soundings and safe anchorages. I took great satisfaction to see that though soundings and shoals (and of course the look of the city) had certainly changed in 200 years, they found Beaufort's sketch instantly recognizable. Then Captain Giunta went to a wide, flat drawer and pulled out the ship's copies of the charts of the Plata, though he admitted he almost never used them. "It's all here," he said, gesturing to the radar showing the ship's depth and the computer system programmed with all the navigational data the ship could need. With the information on channels, soundings, buoys, and landmarks there, he barely needed to steer the boat. Worse than that, on all his charts: not one single sketch. The familiar bird's-eye plans in pale yellows and blues, but nothing to satisfy your eye, to show you what the place looked like. No matter how connected this computerized observation system was with the observations of Beaufort—how completely based it was on the data first gathered then—I felt somewhat disappointed.

So I looked out the window, and there, as the pilot pointed, I saw Montevideo coming. First it was a smudge on the horizon, then a line, then a line with a bump on it that was the mountain—Montevideo, the mountain with a view—and then, almost before my eyes, the harbor that Beaufort had drawn. The mountain on the left, the boats in the middle—though steel cargo ships now, not frigates—and on the other side the rise of the town, from the docks to, I was delighted to recognize, the twin square towers of the cathedral, still just where Beaufort had left them. Focused completely on the approaching shoreline, I made sketches as quickly as I could, identifying each by my only instrument, a compass I had procured in a Buenos Aires shop the day before. The ferry moved so quickly that my perspective constantly changed, and I had nothing like the time Beaufort would have had, anchored off the harbor, to observe and become familiar with the view. Still, we were approaching from almost due west, the direction from which Beaufort's drawing showed the harbor. And on that level I achieved my aim—I came to the harbor just as he had, with sketchbook in hand. I sketched and sketched, stopping only when the captain stepped to a control at the side of the ship and slipped us into the dock.

Those sketches were, in a way, what I had come for. In his sketch, Beaufort turned his surveying data of Montevideo into something real, something human. It wasn't a boundary line, a bird's-eye view he could only imagine; it wasn't a list of sounding depths. It was *Montevideo,* in the way another person could perceive it. A modern-

Beaufort's "View of the Harbour of Monte Video. The Mount and Light house on the left, the Town and Citadel on the right." Beaufort drew it "from the outer Anchorage M" on the chart. Drawn in 1807 and published in 1808.

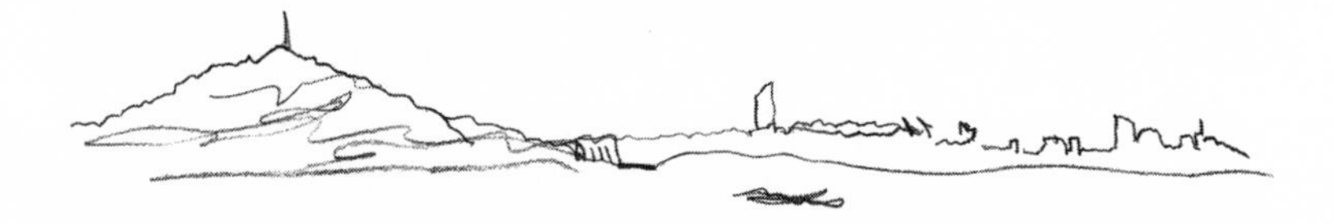

Montevideo in December 2002, drawn by the author, from the cheap seats on the ferry, somewhere around what was once outer Anchorage M.

day seamanship manual I once consulted urged the sailor sketching a shoreline to increase the vertical scale by 50 percent, in order to emphasize landmarks, "giving clarity for the identification of features." In my own sketches I did this without thinking. As I sketched, my friend in the pilot house with me took photographs of the harbor as we approached, and in the photos it looks like a green curb, like any undistinguishable far-off coastline in any photo. Not in my sketches, though; there it's instantly recognizable as Montevideo, the harbor, the city, and the mountain. That is, my own eye, the pencil in my hand, could catch what a camera, an instrument seemingly more exact, could not. It felt powerful, and when I looked over the sketches later the feeling was still there.

As I gathered my pencils and maps, I noticed on the captain's table a pair of binoculars. I pointed and raised my eyebrows. He shrugged. "You know, to make sure. Sometimes fog, sometimes rain, sometimes clouds," he said. "Sometimes you have to look out the window." I felt something like relief.

LOOKING OUT THE WINDOW—or anyway standing on deck looking at the harbor—was all Beaufort would have had at the turn of the nineteenth century.

But that was a great time for mapmakers. Voyages like Cook's in the late 1700s kept turning up more land, and if you were British, your country kept grabbing more of it, for trade and through military means. Britain had gained control of India and, despite its loss of the American colonies, solidified its hold on Canada. Its territories in Africa grew too, as did islands from Grenada to the Falklands to Manila—and every territory obtained, whether for administration or trade, was a new coastline to be mapped. For another, the advances in instrumentation that would make accurate and useful maps possible had all come in the previous half-century or so. For a man possessed of a rage to take bearings and make charts and plans, the world was, almost literally, a blank slate.

First, think about land. As an example, consider that when Beaufort was born, Captain Cook was sailing around in the Southern Ocean looking for a continent that people were pretty sure was hanging around down there—it was still referred to as "Terra Australis," and nobody knew exactly what, where, or how big it was, though people like Alexander Dalrymple, who became the Admiralty's first hydrographer, thought it ought to be pretty big because obviously it had to balance out all those continents in the Northern Hemisphere. It had been seen on Cook's first voyage, but on his second, from 1772 to 1775, he was trying to get a really good idea of what the story was. Of Antarctica, of course, people barely knew a thing.

Which isn't to say that people had such great maps of the places they were familiar with, either. The first trustworthy map of Ireland wasn't made until Francis Beaufort's father got around to it in 1792, but that scarcely put Ireland at the rear of the pack; the first topographical map of France came a year later; the British Ordnance Survey hadn't even been launched until 1791. The Prime Meridian—the ground zero from which all measurements of longitude were made—was still a matter of great international debate and would not be finally agreed upon until 1884. (Even in 1881 there were at least *fourteen different* Prime Meridians in use.) The Earth has seven continents; when Francis Beaufort was born, people had a very good idea of about four and a half of them, and that's if you're being generous about conceptions of Africa, Asia, and South America. Mapmakers had only recently stopped filling the uncomfortably empty spaces of Africa with the imaginary kingdom of Prester John. If you're drawn to making charts and maps, places on the globe where pictures of sea serpents constitute the state of the art are the answer to a prayer.

The Story of Maps, one of the standard descriptions of the development of cartography, describes the situation simply: "All that map publishers lacked was good maps to copy. Nearly all of Europe and parts of Asia and Africa had been surveyed after a fashion, but all too few of the standard maps were based on trigonometric surveys.... European monarchs had staked out possessions in foreign lands, yet not one of them could locate what he claimed."

Which is where Beaufort and guys like him come in.

THEN THERE WERE INSTRUMENTS. To make a useful map you need to know two things: exactly where you are, and where everything else is in relation to that point.

Today, to solve that problem all you need to do is tap a button on a Global Positioning System receiver and you've got your latitude and longitude, usually within twenty feet or so, plus a map of the surrounding area. You can get everything you need at Target. All the navigational equipment I saw on the ferry *Juan Patricio* was just more highly accurate versions of that same technology. Want to know how far away something else is, or make a map, and need more accuracy? Get a modern surveyor's transit that uses an infrared beam to give you an almost preposterously accurate reading of distance and elevation. If it's just altitude you want, there's a reasonable chance you can get that from a fifty-dollar *watch.*

Not for Beaufort. Finding latitude and longitude was work that took hours every day. The best computer was a slide rule. Latitude was by far the simpler to find, and had been for centuries—you just needed to measure the height of the North Star from the horizon in degrees, and that was your latitude. You could get the same result by measuring the height of or relationship among other celestial objects—like the sun and moon—at particular times or points in their paths, and comparing them with the thick volumes of tables that every ship carried.

To take these readings Beaufort would have been able to use a quadrant, which looked very much like the sextant that any modern sailor would recognize. Not long before, sailors would have had to use a backstaff or cross-staff or even an astrolabe to take celestial measurements. A backstaff measured the height of the sun by projecting the shadow of a pointer onto a curved scale in a kind of sundial arrangement. The cross-staff was even more primitive, with the sailor sighting along a pole and adjusting a crosspiece until it fit exactly between the horizon and the celestial object being sighted. Not so bad for stars, but if you were sighting the sun it tended to make you blind.

Fortunately, around 1731, in one of those Newton-and-Leibniz-simultaneously-developing-the-calculus situations, the sextant was invented by John Hadley in England and by an American named Godfrey. After Isaac Newton died, it was discovered that he too had a plan for a sextant. The sextant uses the effect generated by mirrors facing each other to bring the horizon and the measured celestial body together. You look through a scope at a half-silvered mirror through which you can see the horizon; you adjust another mirror to reflect the sun or star you're measuring onto the half-silvered mirror, and once they're together you read the scale on the instrument. The new quadrant—called a Hadley by sailors—vastly improved the accuracy of readings.

It was developed as part of the tremendous push in the eighteenth century to find a way to accurately measure longitude. The other, and eventually highly more useful, method (as described in *Longitude,* by Dava Sobel) was the use of the marine chronometer developed by John Harrison to compare local time with the time at a chosen spot—in England the spot was Greenwich—and thus determine longitude with a degree of accuracy depending on that of the chronometer and the sextant, or whatever other celestial means sailors used to determine local time.

Beaufort was born just two years after John Harrison received his prize from the Board of Longitude for the development of his chronometer. The chronometer first started appearing on ships around the time of Cook's voyages, and the *Vansittart* had one. So for his charting, Beaufort had the best equipment in history—he was part of the very first generation of navigators who could figure out exactly where they were.

Chartmaking tools had significantly improved as well. For measuring angles, Beaufort would have used either his quadrant—turned on its side for horizontal angles—or a theodolite, a sighting instrument with a spirit level and crosshairs, mounted on two plates, one of which rotated vertically and the other horizontally. With vernier scales for exact measurements, a theodolite could measure the angle between two landmarks with great precision. Surveying is all about trigonometry and angles, so a good theodolite or quadrant is essen-

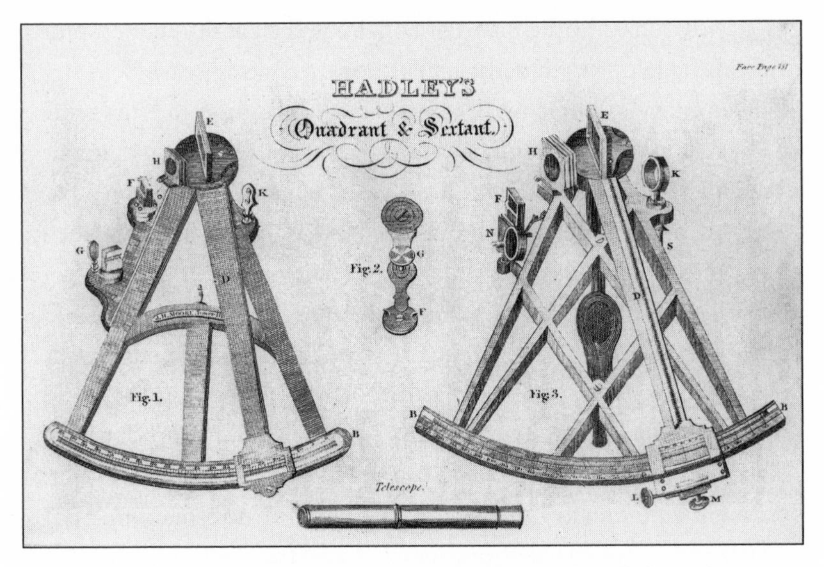

From Moore's *The Practical Navigator,* an illustration of Hadley's quadrant and sextant, nearly a century after their invention.
From *The Practical Navigator,* page 151. Graduate Library, University of Michigan.

tial. The theodolite had been around since the 1500s, but only in the 1720s had someone thought to put a telescope on it, which radically improved accuracy.

Another invention, the year before Beaufort's birth, had an even greater impact on surveying accuracy. In 1773, Jesse Ramsden of England invented the Dividing Engine, a machine that could create measuring scales of tremendous accuracy, using wheels and gears to mechanically divide, say, a single degree on the circular scale of a compass, theodolite, or quadrant. Before that, an instrument was only as good as the craftsman who had, by hand, tapped the gradation of the instrument's scales. Unlike Hadley's new quadrant or the telescopic theodolite, Ramsden's Dividing Engine didn't improve only its own measurements. It made every instrument more accurate. The invention was judged so important, in fact, that Ramsden shared in the reward money disbursed by the Board of Longitude (he got 615 pounds, provided he taught ten other instrument makers how to use it).

The other instruments Beaufort would have used for surveying hadn't changed significantly in millennia. The plane table, for example, had been in use since at least Roman times. It's a table on a tripod that the surveyor sets up directly over a point on his chart, then uses a sight called an alidade (basically that's two slits in pieces of metal, like the two sights on a gun) to locate a line to a landmark and put the line on his chart. Sighting the same landmark from two ends of a measured base—say a bluff or an anchored ship, from either end of a one-mile stretch of coastline—yields not only an accurately placed landmark on the chart, where the two lines intersect, but, through trigonometry, a staggering amount of information about that landmark's distance from the base and from other landmarks. Among that information is the exact length of each side of the triangle, so the other sides of the triangle can thus be used as bases for further triangulation. Surveyors say that with a single measured base of sufficient accuracy, you can triangulate a survey of an entire country.

To measure that baseline on shore, Beaufort would use a standard surveyor's chain; if the baseline was a distance traversed by the ship, he'd use the ship's usual log line (nothing more than a rope let out behind the ship; a log tied to the end remained stationary, and the length of rope paid out in a particular time gave the ship's speed and distance traveled), though for longer distances he would probably take sextant readings from opposite ends of the line the ship traveled.

To measure the depth of the channel, sailors for millennia had been using a lead line for taking soundings, though by Beaufort's time the weight on the bottom of the line would have been loaded with paraffin—samples of the sea bottom would stick in the soft paraffin and come to the surface with the lead. That way, as well as finding out the depth of the channel the ship was in, the sailor would get qualitative information: Was it sandy? Filled with shells? Gray, brown, black? As observations were built up over the years, guides to harbors would include not only descriptions of landmarks but coordinating landmarks on the bottom of the sea to help the sailors know where they were. (For example, Moore's *Practical Navigator,* a standard work of Beaufort's time, includes a table of the "Quality of the Soundings" captains can expect on their way into the English Channel, given in order of distance from the French port of Ushant: "52 Leagues: Fine grey sand, mixed with black . . . 32 Leagues: Sand mixed with gravel, shells, and small cornets.").

This was what made the world Beaufort lived in so different from ours, and what begins to show the raw material of the wind scale I eventually stumbled across in the dictionary. Even the quality of sand gathered by a lead line provided information, and it required attention. In Beaufort's world, *nothing didn't count.* And in fact Moore's book yields plenty of other information about how Beaufort would have done his work, about how everything counted. It describes, for example, how a sailor who takes a bearing on a mountain and then sails a known distance—as could be measured using the ship's log line—directly away from the mountain, could then take another bearing and determine the height of the mountain as well as, once that was determined, the ship's total distance from it. It's simple use of the trigonometry now taught in high school or college, but—*it's the use of that trigonometry.* Beaufort's world was a place where people could figure things out, and they used tools like math to do it. Trigonometry wasn't something to be endured; it was something that helped sailors get where they were going, and make sure they knew how to get there again.

It wasn't always hard math, either. Moore's gives a simple explanation of how to use the curvature of the Earth to determine distance from an object of known height: "To the earth's semi-diameter add the height of the eye, multiply the sum by the height, then the square root of the product is the distance at which an object on the surface of the water can be seen by an eye so elevated." (But a busy sailor needn't bother—the book's Table XXIX provides it.) So a sailor approaching land, who knows of a lighthouse whose top is 600 feet above sea level, can consult the table and learn that when he sees the light first peeping over the horizon, he's about twenty-six miles away. Or consider a merchant on shore watching for his ship to arrive. When his telescope just enables him to see the vane of the ship whose tallest mast is 140 feet above the surface, he can figure that the ship is 12.59 miles off, at least if he's lying prone with his face on the sand. Otherwise he ought to add 2.6 more miles away for the extra distance he can see by being about six feet tall.

It's amazing—and of course it clarifies how improvements in instrument accuracy caused vast improvements in measurement. A common surveying technique was a running survey, during which the crew would never set foot on shore. Using the quadrant to take horizontal angles and the log line to determine distances, the surveyor could use simple instruments and trigonometry—and a steady sketching eye—to create an accurate and detailed chart.

Everything counted—and there was always another way. Perhaps a captain didn't know a distance and couldn't trust his log line, which was, after all, only as accurate as the sailor paying it out and either tipping an hourglass over and over or repeating a certain phrase known to last a certain amount of time. In that case, the captain could use the pendulum. Alexander Dalrymple recommended sound as an alternate way to measure a base. A thread of a little less than a foot with a weight—Dalrymple suggested a bullet—tied to the end would make one swing in the amount of time sound would travel a tenth of a mile. Thus a surveyor on a ship could watch a confederate fire a gun on shore (or on another boat), count the swings between the time he saw the flash and heard the report, and have an excellent estimate of the distance between the ship and the gun. That's the accurate baseline, and the rest is just math.

What is extraordinary to modern eyes about all this work is its very physicality. Surveying is observational work, and the instruments the surveyors used were aids to, rather than replacements for, observation. Any information the observer could get was good information, and it counted—it helped the overall enterprise, which was to understand what the world looked like, and to communicate that most effectively. Beaufort and others like him were trying, for the first time in history, to find a way to say "You are here" and to have *here* mean not *around here* or *pretty much here* but right stinking *here,* and with the advances in instrumentation they were able to *do* it, and when you realize that, you can begin to understand Beaufort's madness for taking his bearings. He was a guy who wanted to know where he was and where he was going, and he lived in a world where for the first time if you had a small chest full of cool equipment, some almanacs, and the kind of education you could get from a nutty but motivated father and some helpful sea captains, you could really know that.

I found it thrilling. People were still figuring out where the continents were; they were still figuring out things like magnetic compass variation and how the tides worked. And when you could, you put down on paper what you could figure out.

FINALLY, KEEP IN MIND, TOO, that mapping was a highly independent enterprise. The French Dépôt des Cartes et Plans, the world's first organized hydrography department, had not been established until 1720, and other national hydrography groups followed through the century, mandated to organize charts and systematize marine surveying. Before then, charts were simply made when they were needed—by both merchant and naval ships—and, more important, were jealously guarded. As exploration continued and new land was discovered, trade routes were established. Accurate charts meant money—if one company knew where the shoals and hazards were in places like India and Africa and Indonesia, its traders would less likely suffer shipwreck and misadventure and thus more likely return home laden with either the goods it had sent treasure to purchase or the treasure for which it had sold goods.

Worldwide shipping traffic was expanding exponentially, and the merchant class was at the core of it. Before becoming Hydrographer to the Admiralty, Alexander Dalrymple had been hydrographer to the British East India Company (and, as I later learned, played a central role in the development of the wind scale). Captains of any ship often—but by no means always—deposited their new charts with the Admiralty, but the charts were often unreliable and poorly organized. When charts were made available, they were published privately and could be a source of income for the surveyor, who retained copyright, though they were widely plagiarized, a process through which further error crept in. All the plagiarizing and error further increased the value of a truly accurate chart. In all, the system was at best disorganized and more often chaotic.

And charts weren't often shared among nations. In the 1700s, when lightly detailed sea atlases were the maps standard aboard most ships, British sailors complained that Dutch ships had more-accurate charts of the British coastline than they had themselves, and charts were much coveted. During the Napoleonic Wars, when the British captured the French ship *Hougly* off St. Helena, the great booty obtained from the ship was her charts. Dalrymple's hydrographic office copied the charts—though it's worth noting he then returned the charts to the French. Even at war, withholding sea charts would have been considered inhumane: a ship's charts were a vital part of its brain—its memory. In fact, during the Napoleonic Wars, the British had greater losses—losses *eight times* as great—from shipwreck caused by poor charts (and, of course, storms) than from enemy action. Accurate sea charts were not just an outgrowth of exploration and improved instrumentation. Accurate sea charts were what the world desperately needed.

THUS THE QUALITY OF BEAUFORT'S CHART of Montevideo helped make his career in the Admiralty. So, with Beaufort's maps and drawings in my head and my sketchbook in my hand, I left the ferry and entered Montevideo. With a portside market specializing in meats and the gridwork of streets unchanged since Beaufort's day, Montevideo feels like the combination of South America and Europe that it is. Fragments of the ancient city walls crumble here and there, a crystal-clear sun beats down on street after street of flat-roofed Spanish neoclassical homes and colonnaded public buildings, all of which have seen happier days.

I was delighted to find that once I had updated street names, my little copy of the map Francis Beaufort drew "to keep the devil out of [his] mind" was far more useful than the one in my travel guide. The street pattern remained the same; a city square skewed out of alignment that had held the governor's house and "Publick offices" in Beaufort's time was now the Plaza Zabala and contained a statue of a man on a horse, but it still had the same off-kilter skew. Beaufort's was both larger and less cluttered, so I used as my guide a map two hundred years old. I found a barracks that in Beaufort's day housed troops and sketched that, invited inside for a better view by a friendly woman who lives now in a partially ruined outlying house, a curtain across the door, a clothesline amid the rubble.

At the very end of Beaufort's map, I sketched the top of the old gate to the city, still standing, supported by a granite wall, at the edge of a

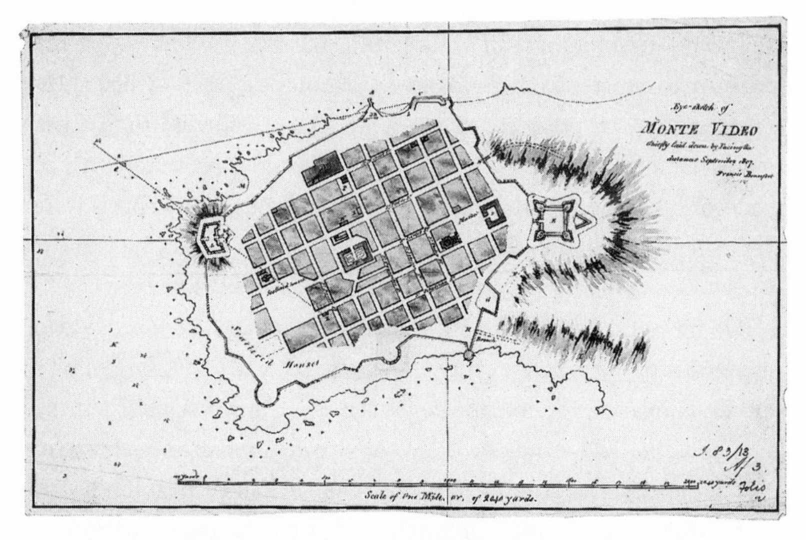

"Eye-sketch of Monte Video chiefly laid down by pacing the distances September 1807," by Francis Beaufort. When I was in Montevideo, this 195-year-old map turned out to be more useful than the maps provided by travel books or tourist agencies.

modern plaza, and using nothing more than a compass, I made a light attempt to map one of the Montevideo squares on which sat the church Beaufort had included. Lacking a sextant or theodolite, I couldn't trigonometrically measure the church's height. I considered the old shadow trick, where you poke a stick in the ground, compare the length of its shadow with its height, then measure the shadow of the church, but I realized I was missing the point. What I loved about Beaufort's map wasn't just its clear edges, scale, and legend, identifying churches and theaters, houses and markets. What I loved was its simple existence—that Beaufort, forced to spend a month in a place, occupied his time by observing, by representing what he observed, and by honing that representation.

I thought of a passage I had seen in Moore's *Practical Navigator.*

That handbook, which might have been on any ship of its day, contains information on more than just sextant use and trigonometry. Along with the tables and math problems, the geometry and astronomy, were general discussions of things you might encounter aboard ship: "The Method of Exercising Merchant-Ships' Companies for War," with detailed suggestions for, say, how many men ought to work each gun; "On steering a Ship that has lost her rudder"; "Directions for the Restoration of the Drowned, those suspended by the Cord, intense Cold, or tremendous Lightning," and so forth. Best, though, is "On saving Lives from a Ship lost on a Lee-shore," where a ship might have to save its sailors from a wreck in which their ship was pushed by the wind onto rocks. Those able to offer assistance must do so—it "must be allowed to be one of the greatest acts of humanity," Moore says—but the dangers are great both to those aboard the wrecked ship and those offering assistance from shore.

Moore suggests the use of a Flying Storm Kite, "as may be easily and readily made on board any wrecked vessel." The idea is simple: Use some sailcloth and a couple pieces of wood to make a standard kite (of about eight feet in height), then use the kite to carry ashore a grappling hook attached to a rope aboard ship. When the hook is over the shore, let the kite go free, and the increasing weight of line will bring it down, at which point it can anchor ashore and connect the shipwrecked sailors to terra firma.

It's almost breathtaking.

Again, the solution shows the ingenuity by which Beaufort seemed to thrive—use everything, ignore nothing. If your boat is wrecked, use the pieces to save yourself. If the wind drives you on shore, find a way to use the wind to save you.

The wind.

I touristed around Montevideo and sketched some more buildings and walls, fishermen and fountains, but I had done what I came to do: I had walked in the footsteps of Sir Francis, I had observed what I could, and I had tried to think the way he would.

I hated to get back onto that enclosed ferry. Not that I minded leaving Montevideo, though I liked the town. But I knew that the way to Sir Francis Beaufort was out in the wind.

CHAPTER 2

In Search of the Wind

This diagram of the sails and rigging of a ship of Beaufort's day comes from Moore's *The Practical Navigator.* (The legend, supplying names of sails, yards, masts, and rigging, appears in Appendix B.)

From *The Practical Navigator*, p. 349. Graduate Library, University of Michigan.

YOU NEED TO KNOW one fundamental fact about wind in the day of Sir Francis Beaufort, the heyday of sail. That fact is simple and, when you think about it, obvious: "The wind," a boatswain named Glen Hyman said to me, "was oil. The world *ran* on wind." We think of solar energy as something caught by a generator or a panel of cells, turned into electricity, and sent through a wire and used to power another machine that does work—lights or a radio or a water heater.

For Beaufort and the people of the nineteenth century, solar energy didn't need translating technologies. It was there, ready to do mechanical work, to provide power. The Earth itself was a perpetual-motion machine, and the motion it provided perpetually was wind.

I started meeting boatswains because I figured the best way to get my mind around Beaufort and his scale was to go onto a ship, a tall ship, of the type he'd have sailed. And Glen's point proved itself to me before I even met him. Before I even knew what a boatswain was.

IN A SNAPSHOT, I STAND in front of a green chalkboard, on which is written in capital letters a foot high, "*EUROPA* IS HERE!"

At the bottom of the board, in much smaller characters, the writer has scrawled, "RTT is still 400 NM Away!!" And then, in the smallest letters on the board, "But still coming!"

Alas, the *R. Tucker Thompson* never did make it to Port Alberni, the tiny port on the west coast of Vancouver Island whose marine festival, centered on the visit of those two tall ships, forms the background of my photo. She remained becalmed in the North Pacific—just as the *Europa* had been—long enough that her schedule overtook her, and she didn't catch up with the *Europa* until a week later, in Port Richmond, just south of the city of Vancouver. That's all getting ahead of the story, but the point is, long before I boarded the *Europa,* on which I hoped to learn about the wind and the Beaufort Scale firsthand, I learned fact number one about the wind: Don't make plans that depend on it. The wind has plans of its own.

The *Europa* is a bark, which means it has three masts, two square-rigged and one rigged fore-and-aft—that is, the front two masts have square sails perpendicular to the ship, like the ones you think of on a pirate ship, and the rear has triangular sails parallel to the ship, like a child's drawing of a sailboat. That's already a store of knowledge about which I hadn't the first clue before I decided to join the *Europa* for the briefest week-long sail, and it hardly scratches the surface of what I learned.

I didn't find the *Europa* as much as the *Europa* found me. A friend, Todd, had done a good deal of sailing on tall ships; he suggested I board the *Europa,* and we met at a Vietnamese restaurant so he could tell me about it. He took out a yellow legal pad and started sketching: He showed me the six sails I'd likely encounter on a ship (a ship, it turns out, is a craft with at least three masts, all of which are square-rigged, though few will quibble with you if you include a bark in the category). A mast on the *Europa* would have six sails on it, starting with the course, at the bottom, and climbing through upper and lower topsails and upper and lower topgallants to the royal, or sky-sail, at the top.

We went through the difference between sheets and shrouds (I would trim sails with sheets; shrouds are the vertical members of the spiderwebby-looking rigging that I could expect to clamber up. I assumed I would surely do so barefoot, with a knife clenched in my teeth). He flipped around terms like *abaft* and *abeam,* which sounded so ridiculously salty that I blushed when saying them; and, finally, he impressed on me the likelihood that on the *Europa* I would see people taking weather readings that would include, Todd assured me, Beaufort Scale estimates of the wind.

I was all for it. I was starting to research the scale in earnest, and for several months I had been focused entirely on the insides of libraries. My brain was beginning to hurt, so Todd's suggestion was exactly right. The place to become familiar with Beaufort's scale was the place Beaufort spent his time, and that was on deck.

The *Europa* would be sailing across the North Pacific in July, and I could join her in early August when she reached Vancouver Island. I flew out a few days before she was due, thus missing the flurry of e-mails from the central office in the Netherlands urging me to wait awhile, since the ship was making a slow passage due to especially poor winds.

It turned out that the *Europa,* with plenty of fuel for the engines she used when the wind was poor, managed to make Port Alberni just in time for the festival. The *R. Tucker Thompson,* a much smaller ship, had to budget more fuel for desalination to keep her crew in fresh water; she thus truly relied on the wind for motion, and the wind let her down.

I was already thinking about things in an entirely new way, and that only intensified the moment I stepped over a low rail and got on the ship. I had scarcely announced myself nervously to the first mate, a hulking, stern-looking blond man named Seth, before the ship absorbed me. A smiling guy named Glen saw me standing around looking lost. He introduced himself as the boatswain—pronounced bo'sun—and threw me a length of rope. He asked me to coil it, quickly showing me how to twist the rope a half turn with each coil so that it hung neatly ("It has a memory," he said. "Trust it."), and to hang it on a specific peg in the boat's wooden side—a half-hitch would do, he said: "You know a half-hitch, don't you?" I said I did, assuming my fingers would remember from my days as a Boy Scout. They did. It felt like a dream.

I don't think that feeling changed for a single moment of the rest of my trip on the *Europa.* Once a simple light ship, built in 1911 as a kind of floating lighthouse in a German river, *Europa* had been refitted around 1990 as a bark. She is 150 feet long from stem to stern, though you can add another 35 feet for the bowsprit, the mast pointing off the front. Her rig height is 109 feet, which puts you a good 120 feet above the sea if you climb to the top of the mainmast. Her deck is 24 feet wide at the beam, with a steel hull painted white, the obligatory figurehead of a naked lady at the prow. *Europa* now makes her living circling the globe, carrying well-paying passengers—up to fifty, but there were only seven of us when I was aboard, most of whom had paid more than $100 a day to sit glumly for a month on the calm Pacific passage. It also functions as the training vessel for the Dutch Enkhuizen Nautical College, which meant that the crew of fourteen comprised largely young Dutch sailors learning to work a square-rigger, though she carried trainees from England, New

Zealand, and Japan as well. She stopped at Tall Ship festivals and was an event wherever she came into port.

She was perfect for my purposes. First, I appreciated her lack of pretense. In search of the origins of a piece of writing that struck me with its straightforwardness and simplicity, the last thing I wanted was to suspend disbelief. There wasn't a single thing about the *Europa* that smacked of reenactment or let's pretend: The *Europa* was a working ship, with hot and cold water, restrooms, and outstanding food. But, second, she is one of the few tall ships sailing that carries a full set of sails—that is, she has not only all her studdingsails (the sails put to the side of the usual sails when the ship is running directly before a weak wind) but the royals and split topgallants that make her, from the deck up, almost a perfect sailing replica of a frigate, the standard three-masted ship of the eighteenth and nineteenth centuries. The exact type of ship on which Francis Beaufort mostly sailed.

I threw my bag in a six-bunk room—the only porthole, just at wave height, was in the bathroom, so with the doors closed it's pitch black. I was distressed to notice that the hum of the engines could be heard in the darkness, even in port—but I didn't stay in the cabin long. Focused on the Beaufort Scale, I applied myself first to joining the captain, Robert Vos, a rail-thin Dutchman with a two-day growth of beard, a shock of sandy hair, and a hand-rolled cigarette perpetually hanging from his lips. He spoke fluent English and had a kind of sarcastic confidence that commanded the loyalty of his crew. If you sent a request to central casting for a modern sea captain, Vos is who they might send.

"I look over the side," he said when I asked him how he applied the Beaufort Scale. He glanced about, looking at the sea, the waves, the birds, and said "This is . . . Beaufort 2." He shrugged and raised an eyebrow. The other numbers were just as simple. He looked at, mostly, the state of the sea—how high the waves were, what the surface looked like—and he wrote it down. The lower numbers didn't require much attention, since he spent so much time on the water; he could estimate them all without much thought. As for the high numbers, "If it's 10, 11, 12, who cares," he said. "It's just surviving anyway." He showed me where he wrote his estimate down in his log, every four hours, along with wind direction and his course. To Vos, the Beaufort Scale was nothing more than a tool, not a big part of his life.

The wind, though, was. I said I assumed that crossing the Pacific largely under steam rather than sail would have made for an easy trip, since he didn't have to manage the crew as they changed sails and so forth. He looked at me like I was from Venus.

"It's better with the sails than with the engine," he said. "That's the power. The engine is auxiliary." The ship was designed to use the wind, and everything else was just a modern way of making the best of a bad situation. More masts break from no wind than from wind, for example: When a swell rolls the ship, if there's no wind bellying out the sails, "the whole rigging bangs against the mast," Vos said. He shook his head. "Banging . . . not so good." As for whether a wind doesn't put more strain on the mast, the wind pushes the mast forward, where it's supported by the entire complex system of shrouds and rigging bracing each mast from the back. "It's *built* for that," Vos said. "But no wind . . ." he shrugged.

Vos talked a lot about not having wind, since the ship seemed to be bringing its calm with it. And after a day of visits by tourists in Port Alberni, the *Europa* raised sail to leave the harbor—it's the decent thing to do—but the sails sagged like bedsheets and the ship steamed away.

THEN AGAIN, CALM WASN'T SO BAD EITHER. I had seen a film about the crew of one of the last square-riggers sailing—square-riggers sailed well into the twentieth century; until the completion of the Panama Canal in 1914, large square-riggers still beat steamships for trips around the Straits of Magellan, since there were few convenient coal stops in South America. The film showed sailors on footropes out at the ends of yardarms above a roiling sea, and it looked kind of scary. Not only that, if you read about wind, you end up reading—you can't avoid it—about shipwrecks.

In the eighteenth and nineteenth centuries, a huge literature about shipwrecks developed: books like *REMARKABLE SHIPWRECKS—A Collection of Interesting Accounts of NAVAL DISASTERS with many particulars . . . together with an Account of the Deliverance of Survivors Selected from Authentic Sources,* from 1813. This kind of book abounded—it was the Discovery Channel of its time, only instead of it being Shark Week every time you turned around, it was Special on Watery Death. It was entertainment for a time with fewer options than our own—in fact, when the seas were rough, people near ports would gather on bluffs overlooking the ocean to watch the carnage.

All of the shipwrecks in these books seem to have the same elements: an uncertain sense of where they are, and often a haughty captain who refuses to listen to reason. Then there's usually a sudden squall. Follows a shudder, then a crash, then they're sawing off the masts to lighten the ship, and within a paragraph they're swimming to shore to face a killer surf and unpredictable natives.

But one book caught my attention. Written in 1704 by Daniel Defoe, it's simply called *The Storm.*

November 1703 was windy in the south of England.

A warm wind, generally from the southwest, kicked up about the middle of the month, and it blew so steadily and for so long that it filled virtually every port with ships. Trade ships or men-of-war in the Atlantic rocketed east before the wind, arriving in England days ahead of schedule; those wishing to leave were battered back to port in the teeth of the seemingly endless gale. Big ports like Falmouth and

Portsmouth had several hundred sail each; the Thames was filled with ships; smaller ports along the South Downs simply overflowed, and ships had to drop anchor where they could.

Then the real wind showed up.

By November 25, things were literally shaking to pieces. Defoe wrote that he "narrowly escaped the Mischief of a Part of a House, which fell on the Evening of that Day by the violence of the Wind." Things were bad enough that "had not the Great Storm followed so soon, this had pass'd for a great Wind." Even so, though on Friday the twenty-sixth it still blew "exceeding hard," he thought it "not so as that it gave any Apprehension of Danger within Doors." Then about midnight people stopped thinking they were safe anywhere, indoors or out.

People cowered in their houses, for "Bricks, Tiles, and Stones from the Tops of the Houses, flew with such force, and so thick in the streets, that no one thought fit to venture out, tho' their Houses were near demolished within." So many tiles flew from people's roofs during the storm that afterward the price of tile went up 450 percent. Reed for thatched roofs rose even higher, and people had to thresh their corn not for the seed but to get straw to fix their roofs. A year later, Defoe reckoned the products of the entire summer tile-making season would not repair the roofs within a ten-mile radius of London.

All night the wind increased, hitting its peak just before dawn. When it finally slackened, Britons crept from their homes to find, as an observer in Brighton put it, "the Town in general (upon the approach of Day-light) looking as if it had been Bombarded." Said another, "the Houses looked like Skeletons, and an universal Air of Horror seem'd to sit on the Countenances of the People."

Defoe, seeing firsthand the damage in London, put an ad in the London *Gazette* asking people all over England to relate their storm stories so he could collect them in *The Storm:* "Such works of providence are worth recording," one said. Think of it as an eighteenth-century version of the videocassettes and special sections that local news outlets release after any tornado or hurricane.

The details are staggering even today. The lead sheathing of church roofs rolled up slanted eaves like parchment. Windvanes bent double. Elms and oaks blew down in groups of hundreds. Chimneys and steeples crashed through the night. There were the usual tales of inexplicable escape: a family was saved by running outdoors just before their house collapsed; another was saved by failing to run outdoors just before the neighbor's house collapsed. Trees, carts, objects, and people were thrown unimaginable distances. Milkmaids, against advice trying to make their way into town, had pails of milk blown from their heads. Some of the stories seem a little theatrical and nowadays might be considered urban legends: Defoe's correspondents tell of families running into the storm and having their nightcaps ripped from their heads "never to be seen again," of people unable to move into the wind even on their hands and knees. And if all the tales are true, England had before the storm a remarkable number of comely widows just waiting for their roofs to be blown away so they could run through the countryside in their nightclothes, raven tresses blowing wild.

Still, exaggeration can be forgiven; the storm may have been the worst England ever experienced. One correspondent says simply, "to speak of Houses Shatter'd, Corn-ricks and Hovels blown from their Standings, would be endless." At least 120 people died, and tens of thousands of head of livestock drowned or were killed. More than 400 windmills near the coast were lost, some because they spun so rapidly that their gears caught fire, *and in the midst of the tempest they burned down.* So much sea spray blew into the air that up to twenty miles inland, leaves and twigs tasted like salt. In short, "horror and confusion seiz'd upon all. . . . No Pen can describe it, no Tongue can express it, no Thought conceive it, unless some of those who were in the Extremity of it."

And that was just on land.

Among the ships waiting out the weeks-long blow, the horror caused probably the greatest sea disaster England had ever seen. With the ships of the merchant fleet and of the Royal Navy choking every British seaport, the ships had no hope of shelter, and in all between 8,000 and 9,000 men lost their lives during the night. "When daylight appear'd," said a fortunate sailor on the *Dolphin,* "it was a dismal sight to behold the Ships driving up and down one foul of another, without Masts, some sunk, and others upon the Rocks, the wind blowing so hard, with the Thunder, Lightning, and Rain, that on Deck a Man could not Stand without Holding." Miles Norcliffe, on the *Shrewsberry,* gave an account of his own ship "run mighty fierce backwards" before its sheet-anchor (the ship's heaviest) finally held, at which point he had the misfortune to see an admiral, "that was next to us, and all the rest of his Men, how they climed up the main Mast, hundreds at a time crying out for help, and thinking to save their Lives, and in the twinkling of an Eye were drown'd."

Dismasted ships were blown so far up the English Channel that they had no idea where they were. The crew of one ship, finally reaching calm offshore of a small harbor, saw a boat leave the harbor and row toward them. When the oarsmen spoke, the sailors could not understand. Natural enough, since they were speaking Norwegian—the ship had been blown north to the coast of Norway.

But the most memorable loss occurred due south of Portsmouth, at England's most important beacon, the Eddystone Lighthouse. One Henry Winstanley, a somewhat eccentric merchant, had completed it only four years before, after some of his own merchant ships ran aground on the dangerous Eddystone Rocks. Made mostly of wood, the lighthouse was either a wonder or a hazard, depending on whom you asked, and Winstanley, sick of the criticism, had supposedly been heard to wish for a great storm to ride out in his lighthouse and thus prove its safety. Whether because his wish was granted or, more likely, to repair damage the lighthouse suffered during the early days of the long blow, Winstanley rowed the fourteen miles through rough waters to his lighthouse on Friday, November 26. The next morning nothing remained of either the lighthouse or its maker.

It was an unimaginable storm, instantly referred to as the greatest England had ever suffered. And Defoe dedicates his entire first chapter to one central point: Though everybody knew *what* happened, nobody had the slightest idea *why*. The wind blew—it blew like hell—but there knowledge stopped.

I read the chapter a couple times before I realized that wasn't just general introductory chatter. Defoe meant this: Some things in nature are obvious, he claims, plain in the relation of cause and effect. Some, on the other hand, remain mysterious. "Among those Arcana," Defoe says, "the Winds are laid as far back as any. Those Ancient Men of Genius who rifled Nature by the torch-light of Reason even to her very Nudities, have been run aground in this unknown Channel; the Wind has blown out the Candle of Reason, and left them all in the Dark. . . . The deepest Search into the region of Cause and Consequence, has found out just enough to leave the wisest Philosopher in the dark, to bewilder his Head, and drown his Understanding." Why do the winds blow? God only knows: Nature herself is enraged by the question, "and at last, to be rid of you, she confesses the truth, and tells you, 'it is not in Me, you must go Home and ask my Father.'"

What I couldn't shake was that this was the 1700s—Columbus had discovered the New World, Drake and Magellan and countless others had sailed all over the world, powered by the wind, yet they didn't know why the wind blew. I was interested in how the wind worked, and how they perceived it, but they, it seemed, were not. That wasn't important to them. They knew something much more useful: how to work the wind. They had machines for that.

A MACHINE TO WORK THE WIND. I was first stunned by the miracle of the *Europa*'s power the second time I climbed aloft. Aloft, of course, refers to the rigging and yardarms clinging to the ship's three masts. Unlike in Beaufort's day, people going aloft on the *Europa* strap on harnesses they clip to lines for safety, but aloft is aloft. And many of the sailors do indeed climb barefoot—it's a tradition aboard ship, where the deck is worn smooth—though none that I ever saw ever clamped a knife between his teeth.

The first time I went up, I did so as a tourist—it's part of the experience, and I was glad of it, and glad to come down. But in a closed ecosystem like a ship, you're either crew or cargo, there's no middle ground. I had been delighted to find, when Glen threw me that rope, that I still knew a knot or so and could be useful, and from the moment the ship was in motion I was mad to help. The *Europa* is a training ship, so I often gave one of the fourteen crew members the opportunity to find out how well he or she had learned a new skill by teaching it to me. What's more, that constant cycle of learning and teaching was, in some ways, uninterrupted, from Beaufort, and those before him, to me.

I fell in with a trainee named Ruth, a Dutchwoman in her twenties, who had advanced in her shipcraft to become a sort of manager of the foremast crew. I was one of a group of about five, pulling ropes individually or severally on command, learning to use my body weight to help tighten a resistant line, to jump up and pull down while someone else pulled across, to hook a block onto the ship and gather help when we needed a gang to pull one of the biggest lines—a brace, to turn the mast, or a halyard to raise a yard.

When Ruth was asked to climb the foremast and overhaul the buntlines and to take me up and get me to help, she shrugged, and up we went. Buntlines are light lines that run from the foot to the top of a sail, used to prevent them from bellying during furling—think of bunting and you've got the idea. Buntlines can do a bit of their job at the wrong time, though—the weight of the line itself can hold them against the sail, preventing the sail from reaching its full shape with the wind. Thus when the ship is trying to make the most of little wind—a constant condition during my week on the *Europa*—the crew can go aloft to overhaul the lines, meaning to pull extra line up through the tackle and use light string to tie it off, thus leaving slack in front of the sail and transferring the weight of the heavy line to the tackle, allowing the sail to swell free and make as much as it can out of the small puffs of wind on offer. It's borderline make-work, giving sailors something to do in light wind. But for Ruth it was a chance to practice her craft, and for me it was an excuse to go aloft.

So up we went, and we overhauled, from topsails all the way up to the topmast. It felt good to be useful, and it felt great to be so high up in the air on business rather than just hanging around; having work to do made it feel less scary, somehow. But more important, from aloft, as you stand on the footrope beneath a yard, out over the water—clipped on to a line for safety—you can survey the whole ship, from bowsprit to spanker. You can see how the masts are angled to send the ship where the captain wants it to go—a square-rigger can go up to about 70 degrees into the wind, the yards sweeping a good 150 degrees between their farthest points fore and aft. You can see the captain on the bridge, making light adjustments, turning the wheel, and calling to the crew to haul here, release there, and a sail gets another degree of belly in it, or the ship leans just the tiniest bit, and you can feel it: "Ah—he's caught something more."

From up there, that is, you could see the ship working, see it doing its miraculous job of catching the wind and turning that into forward motion, in a chosen direction. I thought about it—about Beaufort and the millions of sailors before and after him, about a machine that used the wind to send you anywhere in the world. Up there, maintaining the machine of the ship and watching it function, I decided that the wheel was probably more important, but after that I couldn't think of a single piece of technology that had affected the world more powerfully than the sailing ship.

When Ruth and I were done, I came down the mast full of this sudden understanding of the ship's mechanics; as usual Glen, the boatswain, was there for comment, nodding and smiling. "Everything on the ship does a job," he said. "Everything has a purpose. Somebody put it there, with some sort of intention. There's a certain amount of real estate, and you have to get everything running fair and true." He looked around the ship—all sails were set, and though they weren't pulling very hard, they were catching the wind. White canvas everywhere, with taut lines connecting the corner of this with a rail on that, sail to yard, block to pinrail. "It's a magical machine," he said. "Magical." Then he went off to some other task.

Although, of course, we were under steam just about the whole time. We had a Tall Ships festival to hit in Vancouver, and one in Seattle after that. A sailing ship on a schedule tends to run its motor.

While I was aboard, the *Europa* made a brief spin around Vancouver Island, making a day stop at Schooner's Cove and ending up in Port Richmond, south of the city of Vancouver. The crew set sail every time there was the smallest breath of wind, but it was mostly for practice and exhibition. When up, the sails rarely caught much wind, and the steady thrum of the engine, auxiliary or not, was our constant companion. As a sailing trip, as an excursion into the wind and the nineteenth-century world of Francis Beaufort, I feared my trip was going to fail.

Still, as time went by, without quite noticing, I began, almost imperceptibly, to get exactly what I came there to get. Just being aboard ship—in a tiny universe, 185 feet long by 25 feet wide by 109 feet high, spread over four decks—changed my perceptions and brought them into a focus that, once again, started to give me at least a small perception of what kind of world had created the Beaufort Scale.

Beaufort's sister wrote that when he was a boy of only five, Beaufort already wanted to be a sailor. It's hardly surprising—being a sailor, to a British boy of the late eighteenth century, was like being a steamboat captain to an American boy of the mid-1800s, an astronaut to one of the 1960s. It was romantic and adventurous and about the coolest thing a person could do. There was danger and uncertainty, travel and status. Plus, with England perpetually at war, a great portion of life aboard ship was devoted to chase—catching, engaging, and seizing enemy ships. It was a sort of organized, legal piracy, and the ship's crew shared in the spoils. Working on one of the ships in the navy of Admiral Nelson, the hero of the Napoleonic Wars, was like playing for the Yankees, or the Dallas Cowboys.

Still, that's only half of it. Life aboard ship could be miserable—when Winston Churchill described naval life as "rum, sodomy, and the lash," it was this navy he was talking about. And even at its best, working a ship of the navy was an uncertain life. You could lose one of those engagements and end up prisoner or crew on an enemy ship, for one thing; for another, storms or battle could kill you and often did. Even something as simple as mail was guesswork at best; if you checked the drop house in port for mail the day before you left, and a ship full of mail for you arrived the next day, it was just too bad—communication was as unpredictable as the ships' schedules (the worst naval battles of the War of 1812, for example, were fought months after the armistice, but before the ships got the news).

At the mercy of wind and weather—and hostile ships, to say nothing of hostile locals when the ships came to port—ships left England scheduled for a voyage of a year or so and returned years later, manned by whoever was on board at that point. Sailors abandoned ships to settle, deserted ships because the naval life was too hard, or were killed in battle; they missed ships that left port without them, or ended up on other ships that sailed with them. Captains who needed crew in foreign ports made the best crew they could with whoever was at hand, so ships commonly ended up with sailors from many countries and continents literally pulling together.

But one of the results of this constant and unpredictable mixing was a standardization of rigging, Vos told me. From ship to ship, from country to country, captains and crew leaders needed to be able to communicate with crews who could easily be made up of men who couldn't even speak the same language. "Especially during a storm," Vos told me. "You can't hear each other; you don't even know who's shouting." Thus, a kind of standardization took place for the rigging of the sails.

For a group of visitors on a day-sail we took from Schooner's Cove, Vos and Seth, the mate, gave a little lesson. Seth first explained how to reorient the masts to the wind. "First you brace, and then you set the sails," he said, using a little balsa wood model of a mast to demonstrate. Bracing was rotating the mast; setting the sails involved pulling the lines that tightened them. "You loosen the stays on one side, then tighten them on the other," he said. "Then you loosen the clews and bunts, haul on the sheet for the lower topsail." Then you continued on, making your way up the mast—you maintained control of the ship by starting with the lower sails first, because the higher you went the stronger the wind was, and the more it blew slightly to the right, unrestricted by the friction of the water.

As for knowing which lines to pull and when, that's where Vos took over.

"There are two principles," Vos said. "First, it's groups of four." Each sail had four types of lines: a halyard for raising the yard; a sheet for lowering the sail; clew lines, connected to the lower corners of the sail;

and buntlines, which went over top and connected to the bottom. The second principle is the lower the sail, the more forward the lines are on the pinrail, the rail running alongside the ship, or at the base of the mast. Finally, the *Europa* is called a portside ship—that is, everything starts at the port side, so the lines for the first sail would be to port—the left—then alternating thereafter. Utterly systematic—and knowing the system could put you in the right place during battle or rough weather, when it might be important for you to be there, whether you were Dutch or British or African.

And it worked—I was an example. By the end of the second day hauling lines on Ruth's foremast, I could distinguish between the Dutch for "topgallant" and "topsail." It was a system—a standardized way of communicating a lot of information in a few words. And it would have been central to Beaufort's life.

IF WIND SEEMS PRIMARY, that only makes sense. Wind is, after all, quite literally the first thing there was. In the beginning, whether in Genesis or before the Big Bang, there was nothing. Then the nothing exploded into itself and blew up into infinite something, moving outward at the speed of light. Then suddenly you've got a starting point, so you've got time, and as you move farther from the starting point you get space. All the stuff moves outward, through the void, leaving time and space in its wake. All that stuff—the universe—is energy moving forward through time. That is, all that material is a species of wind. Energy moving forward through time. Wind has been blowing since the dawn of time, and it'll never stop. The wind is energy made real, the spirit made flesh. Not for nothing do scientists use the name *solar wind* for the particles from the sun that we don't perceive as radiant energy.

That's a little cosmic, but ratchet it down to our more parochial concerns right here on Earth and the same concept completely applies. I took a course in atmospheric physics because I wanted to know. Instead of photons and universal cosmic matter through the vastness of intergalactic space, it's air molecules moving in the Earth's atmosphere, but wind is energy in motion, pure and simple.

Here's how it works. Deep inside the sun, under monstrous pressure, four hydrogen atoms smash together, forming one helium atom. Each hydrogen atom weighs 1.008 atomic mass units, and a helium atom weighs only 4.0026, a good 29-thousandths of an atomic mass unit short. That missing weight is released as energy. It's not much, but for one thing the sun is big, so there are a lot of those explosions. For another, as Einstein figured out, you have to multiply it by the square of the speed of light, which is a lot—so much that it gets to be about 27 million degrees Fahrenheit in there, and eventually that energy radiates its way out in all directions. About one two-billionth of that nuclear energy reaches Earth. That turns out to be plenty.

The fundamental thing that happens then is that the Earth is a sphere, so the sun shines a lot more at the equator than at the poles. Plus, the surface of the Earth is not uniform: Some of it is land, and an awful lot of it is water. That uneven heating of the planet is what eventually causes wind.

Over the equator, the air heats up and expands, taking up more space. Way up at the top of the atmosphere, that air starts to spread out over the cooler air to the north (or to the south, in the Southern Hemisphere), which hasn't expanded as much. Since air has weight, more air increases the pressure on the cooler air at the surface. Since it's a universal law that a gas under higher pressure will flow to a place of lower pressure, right down at the surface, the cooler air under higher pressure flows into the space occupied by the warmer air under lower pressure. That flow is wind. It's just nature's way of trying to equalize the temperature and pressure of the atmosphere.

The same thing happens on a smaller scale every day at the seashore. The land doesn't absorb energy well, so its top level heats up fast and starts radiating that energy right back out at the air. Water, on the other hand, partially reflects the solar energy; what isn't reflected can be absorbed into the water, heating much more deeply, and part of the energy is released through evaporation. Thus it takes way more energy to heat up water than the surface of land.

So the land radiates energy right back into the air, and the air above the ground expands and goes a little higher over land than it does over ocean, which means that while the land is hot—during the day—air from the sea flows toward it. The opposite happens at night. That's your daily sea breeze, and that, in a nutshell, is it.

The air transferring between the hotter equator and the cooler pole (both in the atmosphere and in ocean currents, which are a slower example of the same thing) accounts for the major energy transfer of the planet, the cooler and warmer areas of the atmosphere roiling around like cream in coffee.

You have to throw in the Coriolis force to get the whole thing. Air from the equator heats up, moves to the top of the atmosphere, and spreads out. As it moves—north, say—that upper-atmosphere air spreads out to a latitude whose circumference is less than the equator's, so that air ends up moving, in the direction of the earth's rotation, faster than the Earth is. So it seems to move to its right, eventually straightening out into the jet stream, at around 30 degrees of latitude. At the surface, the cooler air makes its way back toward the equator, with a larger circumference than the latitude where it started, so it too seems to turn to its right—think of the Earth turning beneath it as it flows. Thus in the Northern Hemisphere, way up in the atmosphere, at around 30 degrees latitude you get the jet stream, flowing generally from the west, while at the surface between 30 degrees latitude and the equator, you get consistent easterly winds:

the trade winds, the most important winds in the world. In fact, some claim they are called the "trades" not because they enabled merchant ships to cross the Atlantic but because of their steady track—the root of "trade" is the same as the one for "tread." It's not impossible to believe that "ply a trade" was first said of ships cutting their way through the ocean on steady winds.

North of the trades are the westerlies, in which most of us live; and to truly understand the wind you have to throw in ocean currents and the friction of the Earth and mountain ranges and a million other complicating factors, but if your kid asks you why the wind blows, there's your answer: hot air at the equator; cold air at the pole. A rotating planet. Basic physics—and we pretty much get it by now.

DEFOE DID NOT KNOW ANY OF THIS. His contemporaries didn't know it either, and it wasn't for lack of trying. Only a few years before, people still thought winds were generated by exhalations from the Earth, or came from holes in the ground or the mountains. In 1684, for example, Dr. Martin Lister, writing in the *Philosophical Transactions,* opined that the trade winds were caused by the "constant breath" of seaweed. The regularity of the trade winds across the ocean, compared with the variability of breezes on land, made their differing sources obvious: "The matter of that *Wind,* coming (as we suppose) from the breath of only one *Plant* it must needs make it constant and uniform: Whereas the great variety of *Plants* and *Trees* at land must needs furnish a confused matter of *Winds.*"

That sounds crazy, but the thing is, it comes directly from Aristotle, who provided the first written theory of winds (though the Bible and Homer each numbered four winds, coming from the four corners of the Earth—Homer calls them Boreas, Euros, Zephyros, and Notos, or North, East, West, and South). Aristotle, in his *Meteorologica,* explains that the wind comes from the ground. The sun draws out "exhalations" from the earth, he says. Moist exhalations cause rain, and dry ones cause wind. He disagrees with people like his predecessor, Anaximander, who said wind was simply air in motion. "It is absurd to suppose that the air which surrounds us becomes wind simply by being in motion," Aristotle says—to do so would be like calling any current of water a river simply because it was moving. He also explains that "the winds blow horizontally; for though the exhalation rises vertically, the winds blow round the earth because the whole body of air surrounding the earth follows the motion of the heavens." That is, the rotating celestial sphere that spun around the Earth, carrying with it the stars and so forth, imparted its motion to the wind.

Okay, so Aristotle was wrong in just a few particulars. But at least he was thinking about it. The ancient Greeks thought about that kind of stuff, though—and they thought about it in practical terms. They wanted to understand so they could use the winds better—Homer's names simply identified them by their direction of origin.

You can still see what they had in mind. In the second or first century B.C., in the Roman period in Athens, near the base of the Acropolis, a guy named Andronikos Kyrrhestes built the Tower of the Winds—an eight-sided structure, it functioned as a sundial and supposedly had atop it the world's first windvane. Beneath the vane, though, lay more than simple directions. Each of the eight faces of the wall has on it a relief sculpture of the wind that originates from that direction: Boreas, from the north, was an old man dressed warmly, bearing a conch shell that stood for the noise of the wild northern wind; due south is Notos, a young man lightly wrapped, emptying a jug to show his wind's propensity for rain. The fair west wind, Zephyros, is a lovely young man wearing only a loose mantle filled with flowers. It wasn't just during the second Iraq war when we started to hear about the *turab*—a hot, clammy, and dusty wind that blows through Baghdad—that the winds had names.

In fact, naming specific winds is a worldwide practice, and it's been going on since people were naming anything. In early days proper names identified personified winds with specific properties—strength, humidity, direction. Sailors familiar with them could recognize them by feel, and in fact the first examples of what we call compass roses were wind roses—first with eight points, referring to the eight winds identified in Athens, and thereafter sixteen and finally thirty-two, as there are today. It wasn't until the fifteenth or sixteenth centuries that the terms we use now (north, southwest, east-northeast) were adopted. Until then winds had names, and usually plenty: from the hot, dry Brickfielder in Australia to the Tramontana, blowing pleasantly from the north in the Mediterranean (a stylized *T,* for Tramontana, by the way, is suspected to be the ancestor of the fleur-de-lis we place atop our compasses to signify north nowadays); from the hot Foehn winds that roar downhill in the Alps to Pampero winds that blow north onto the Uruguay coast to the Santa Ana winds in California, which Raymond Chandler says "come down through the mountain passes and curl your hair and make your nerves jump and your skin itch. On nights like that every booze party ends in a fight. Meek little wives feel the edge of the carving knife and study their husbands' necks. Anything can happen. You can even get a full glass of beer at a cocktail lounge." The use of names for hurricanes did not begin until around 1940, but still, when Beaufort eventually gave his winds specific names in 1806, he was hardly ahead of the curve.

Still, Aristotle's teachings on this matter had a profound effect, and Lister (who thought wind came from seaweed) wasn't the only one still under his spell. In 1668, Margaret Lucas Cavendish, Duchess of Newcastle, published her book *Grounds of Natural Philosophy.* "The strongest Winds are made of the grossest Vapours," she says, and then goes on to explain that the nature of a particular wind has to do with how big a part it is of a sort of giant circle. "Concerning the Figurative

Motions of Vapour and Smoak, they are Circles; but of Winds, they are broken Parts of Circular Vapours: for, when the Vaporous Circle is extended beyond its Nature of Vapour, the Circumference of the Circle breaks into perturbed Parts; and if the Parts be small, the wind is, in our perception, sharp, pricking, and piercing; but, if the Parts are not so small, then the wind is strong and pressing."

Hmm.

And whether Defoe was aware of it or not, less than twenty years before the great storm he chronicled, the first of two essential leaps forward in understanding of the winds actually had taken place. In 1686—two years after Lister shared his theory concerning seaweed—Edmund Halley had written up his theory of the cause of the trade winds for the *Philosophical Transactions of the Royal Society.* He was the first person to suggest that the ultimate cause of wind was the heating of the atmosphere by the sun, and in all he came pretty close to the final mark. Unfortunately he decided the trade winds blew generally from the east because the sun rose in the east, heating up the air. As that air rose, cooler air would rush in, creating a following current of air, a sort of general east wind. Halley's article also included what is probably the world's first weather map, in which he sketched the general directions of the trade winds, and he was mostly right—but the theory was lacking.

Then finally, in 1735, George Hadley (the younger brother of John Hadley, who invented the reflecting quadrant), also in the *Philosophical Transactions,* did get it right. He figured out how the motion of the Earth caused the general motion of the winds between the equator and the latitude 30s, and he even guessed there would be prevailing westerlies north of the 30s in the Northern Hemisphere and south of them in the Southern. Anyone living in North America who's ever watched a weather forecast knows this is largely true.

Scientific understanding doesn't spread quickly today, and it wasn't any faster then. In the mid-1700s a French scientist named d'Alembert decided Hadley was completely wrong, and that the winds were actually caused by gravitational forces, especially from the sun and moon. And in 1801 John Capper, an East India Company sea captain, wrote a book called *Observations on Winds and Monsoons,* in which, more than a half-century after Hadley got it right, Capper stuck resolutely with Halley's conclusion about the wind.

But above all, why the wind blew was for the philosophers to worry about. The sailors just needed to work with it.

OR NOT TO WORK WITH IT. The *Europa* spent almost the entire week I was aboard making more wind herself than she found, and we got plenty of practice raising and lowering sails—we'd raise them with every puff of breeze, then lower them because they were actually slowing us down.

Just the same, being on a ship makes the wind your partner in a way I hadn't expected. For one thing, aboard ship, I found the life far closer to those of sailors of another generation than I expected. For another thing, once I had joined the crew, Vos assigned me a watch, which meant I slept in four-hour shifts like everyone else. This focuses your attention wonderfully on the present. "I wonder what month it is sometimes," Glen said. "Or, what meal am I being awakened for?"

Like all members of the crew, I took an overnight shift at the helm, and fortunately for me we did have wind at that time, so in the dark and silence I heard no motor, only the creak of lines and the quiet shake of sail. Seth told me the course—it was 305 degrees, if I'm not mistaken—and though I had an easier time keeping an eye on the internally lit compass than one of Beaufort's men would have had with a candle or lamp, I felt the same wind, the way the ship resisted me if I adjusted too quickly, nestled comfortably against it like a cat if I moved the wheel just right. I could feel when we started to lose the wind, and a sail needed to be adjusted: "When it starts flapping, you tighten it up," Vos told me. "And when I say tight, I mean you need to be able to play the violin with it." I noticed, too, that if I adjusted the wheel too often or too quickly the ship wanted to start a squiggly, back-and-forth course that you could only fix by steering less, not more. A life lesson there.

The few times we did have wind, it was an amazing thing to see the crews naturally split to work the machine: the mizzenmast crew—that's the back one, rigged fore-and-aft, with a sail called the spanker; the mainmast crew; and the foremast crew, where I usually worked. Each crew had a leader, and Vos barely had to shout—Seth, the mate, paced from mast to mast, getting the crews going: "Brace the main," he'd shout, and the chief of that crew would start chanting directions, and her entire crew would grab the lines that rotated the mast, leaning into it, and the latticework above the center of the ship began to twist. "Foremast!" he'd shout, and off we'd go. Then came the individual sails—starting from the bottom and moving up. "Main course!" he'd yell, and a crew would put their backs into the ship's heaviest yard; then it was on to topsails and topgallants. Vos kept the mizzen crew moving, and one by one sheets of white appeared above our heads, luffing for a moment until they were tightened, and then they seemed almost to snap into place, and the ship was racing along, pushed by the wind, powered by nuclear fusion 93 million miles away. Once the sails were set and the lines coiled and stowed, we could wait for further orders—and fly before the wind on this machine.

The sails scarcely flapped, and the only sound was the occasional creak of rope or the ship itself, and the constant hissing rush of the water as we scudded across the ocean. With the ship moving, blowing his hair from his eyes, Vos folded his arms and looked around his ship, just admiring it. Now that it was finally under sail, I asked what would happen if he needed to stop the ship. He shrugged—to get his license he had to do just that. "We could stop within one length of the ship," he said. I was stunned to learn he had that kind of control, and he smiled. "The old guys," he said, "they figured it out. There's not much we can do to make it better."

At the moment, that seemed true. The auxiliary engines were blessedly off; all else was silence. I was reminded of a passage I had run across from *Voyaging with the Wind,* a guide to sailing square-riggers published in 1975: "No throbbing sea-punching propeller thundered behind them, no noisy shaft deadened the sound of wind and sea. . . . No whining old pump disturbed the natural sounds that were the square-rigger's orchestra—a harmony of wind, ship and sea."

I had finally found the wind.

Now: If only there were a way to accurately describe it.

CHAPTER 3

The Beaufort Scale, and Who Wrote It, in a General Way

Hereafter I shall estimate the force of the wind according to the following scale, as nothing can convey a more uncertain idea of wind and weather than the old expressions of moderate and cloudy &c &c.

0	Calm	7	Gentle steady gale
1	Faint air just not calm	8	Moderate gale
2	Light air	9	Brisk gale
3	Light breeze	10	Fresh gale
4	Gentle breeze	11	Hard gale
5	Moderate breeze	12	Hard gale with heavy gusts
6	Fresh breeze	13	Storm

And the weather as follows &c

b	Blue sky	h	Hazy
f	Fair weather	dp	Damp air
d	Dry warm atmosphere	fg	Foggy
s	Sultry	r	Rain
p	Passing clouds	sr.	Small rain
c	Clear, i.e. that is clear hard horizon but not the sky	dr	Drizzling rain
cl	Cloudy	hr	Hard rain
w	Watery sky	sh	Showers
wd	Wild, forked, confused threatening clouds	hsh	Hard showers
dk	Dark heavy atmosphere	st	Settled weather
l	Lightning	sy	Steady breeze
t.	Thunder	sq.	Squally
g	Gloomy dark weather	hsq.	Hard squalls
gr	Greasy threatening appearance	bk.	Black horizon & clouds
		thr.	Threatening appearance

H K F Courses Winds Weather Monday 13th Jany 1806. Woolwich.

[illegible]

The original entry in the journal of Francis Beaufort.
Note that the scale goes to 13.

From the journals of Francis Beaufort, Met Office,
National Meteorological Archive. Used by permission.

SO WE HAVE TO GO BACK to the *Woolwich*, the ship aboard which Sir Francis Beaufort surveyed Montevideo. We have to go back to January 1806, when he had just assumed his command, before he ever left port. As the ship waits for orders we join him in his commander's cabin, taking out his journal—it's a nice brown leatherbound patent notebook, from a company called Williams's, and though Beaufort has carefully labeled it the journal for the *Woolwich,* starting June 1805, he hasn't noticed that he started the book upside down, with the margins askew. An easy mistake to make—margins and bound journals were still a fairly new technology. Anyway, it's no bother.

It's a kind of weather journal, a personal log that Beaufort kept, as many captains did, in addition to the official ship's log. Beaufort, the type of person who wrote everything down, used his log, especially, to record the weather.

Sometime on the twelfth or thirteenth of January 1806—the passage comes after the entry for the twelfth and before the next—Beaufort, in a moment that has become almost mythic in meteorological circles, wrote the following words: "Hereafter I shall estimate the force of the wind according to the following scale, as nothing can convey a more uncertain idea of wind and weather than the old expressions of moderate and cloudy, etc. etc."

He then made a list:

0. Calm
1. Faint air just not calm
2. Light airs
3. Light breeze
4. Gentle breeze
5. Moderate breeze
6. Fresh breeze
7. Gentle steady gale
8. Moderate gale
9. Brisk gale
10. Fresh gale
11. Hard gale
12. Hard gale with heavy gusts
13. Storm

He also determined to list weather conditions by initial—*b* for blue skies, for example, or *r* for rain. It was a simple system, saving space and time and regularizing terminology.

Thus was the Beaufort Scale created.

It's a great moment. I ignored, for a while, a footnote in *Beaufort of the Admiralty* about whether Alexander Dalrymple, the first Hydrographer to the Admiralty, and a fellow named John Smeaton, who had developed an earlier wind scale, weren't part of this process. I loved the idea of Beaufort, the intense sea captain who cherished measurements and bearings, smiling with pride as he wrote down the series of numbers and categories that was to preserve his name.

I encountered Beaufort's journal in the Scott Building of the Met Office archive, in Bracknell, an hour's train ride from London. Now completely absorbed in the wind scale and its creator, I had traveled to England to further investigate its history. The Met Office is the British equivalent of the National Weather Service in the United States, and the Scott Building is its secret hiding place, and it looks it. You'll need a map to find it, in a cul-de-sac with a few other strange buildings where only meteorologists go, half a mile from the main Met Office. In a cinderblock room filled with cabinets and wooden library tables set up for researchers, I asked them nicely and they went into the archives and pulled out the storage box that contains that original journal of Francis Beaufort, and there I was, examining the careful strokes of Beaufort's pen. There's an old IBM card in there, too, folded over lengthwise, that says "Beaufort Scale"—it obviously used to be a label for the journal, which, I could tell from how easily it flopped open to the famous page, was once on display in some glass case in the Met Office. Inside the right front cover of the journal it even says in pencil, "display case in lecture room." Nowadays the journal stays in its box in the back, though there's a photograph of the famous entry on the wall, along with the weird weather objects and maps that constitute the Met Office's display trophies—huge engravings of "frost fairs" held when the Thames froze over in 1683 and 1739, with horses and buggies on the ice; a framed declassified letter from the Met Office to Dwight Eisenhower, dated 21 June 1944, informing him that his decision to invade Normandy on June 6 had been wise, since the weather only got worse thereafter: "mainly force 5 or above," the letter says of the wind. "Thanks," Ike scrawled on the note before returning it. "And thank the Gods of war we went when we did—D.E."

There are old weather instruments—like those glass globes used to focus sunlight on a light-sensitive slip of paper, thus recording the amount of sunlight in a day—hanging around on metal file cabinets, sharing space with models of giant hailstones, stereoscopic pictures of clouds, and what is billed as "Word Processor, Mk. 1," which is an old wooden type-stamping set.

It's like heaven for people who fetishize the weather. Nobody there could explain why the original Beaufort Scale wasn't on display anymore, or when or where it actually had been.

It turns out, of course, that it's actually not such a bad thing that it's not on display. Beaufort stroking his chin, a lightbulb—all right, a candle or a whale-oil hurricane lantern—going off over his head, and suddenly the wind scale springs into being? That makes a great story, and to a degree it's completely true. It's the story you'll find in *Beaufort of the Admiralty,* though in subsequent years a trail of journal articles tells a different one, and you'll also find the different account in a more recent biography of Beaufort, *Gale Force Ten.* Yet Beaufort definitely did sit in his ship, take out his ink and his quill, and make the journal entry. I saw it—it's *there.* Only one error has crept into the mythology, and that is the small matter that Francis Beaufort didn't actually write the Beaufort Scale.

Or he didn't but he kind of did, though really he didn't, though yet again actually he sort of did. Or I guess it depends on how you look at it.

There's compelling evidence that Beaufort actually copied that scale directly out of a pamphlet written by Alexander Dalrymple, then Hydrographer to the Admiralty—and more about that in a bit. But however you look at it, the scale I fell in love with when I found it in the dictionary 180 years later, the scale I loved for its beautifully descriptive language and stunning utility turned out to be, at its conception, this . . . list. This boring list of words. There had to be something more.

IF BEAUFORT HAD MEANT to solve the problem of conveying "the wind and weather in uncertain terms [like] the old expressions of moderate and cloudy, etc. etc.," he hadn't done much of a job of it. Yes, he had put the words in order, and there's value in that, but surely sailors knew that a moderate breeze was stronger than a gentle breeze, that a light air was more wind than a faint air. Moreover, one sailor's gentle breeze might still be another's moderate breeze, one sailor's brisk gale another's fresh gale. Even though this was Beaufort's private journal, he often kept hourly wind readings in it, which meant he was expecting the sailors on the watch to note the wind. He wasn't much closer to his goal of excising uncertainty; in fact, he wasn't any closer at all. If this was the Beaufort Scale, then the Beaufort Scale wasn't much.

Fortunately, sometime in the next year or so, apparently during his trip to Montevideo, Beaufort recognized this. On September 14, 1807—the day before he left Montevideo by sail—he began another journal, and as was to become a habit, he carefully penned on its flyleaf the wind scale he was to use in keeping the weather journal. But he didn't simply copy his previous scale. This time he wrote something far more thorough.

The terms were largely the same, though Beaufort had done some fine-tuning: he had removed "faint airs," apparently figuring that "light airs" got the job done; changed "hard gale with heavy gusts" to the simpler "storm"; and added "hurricane" to the end of his scale,

Soon after copying the original scale into his journal, Beaufort improved it by basing it on the sails of a ship, which he could observe. This updated scale appeared at the front of one of his subsequent journals.

From the journals of Francis Beaufort, Met Office, National Meteorological Archive. Used by permission.

which now stopped at 12. Much more important, Beaufort had this time done what he had set out to—in place of the uncertain method of sailors simply looking around them and estimating the force of the wind based on some sort of general impression, this time he offered something by which they could gauge it.

They could measure the wind by the ship itself. All they had to do was look around them: The ship would tell them the force of the wind.

Of light air, 1 on the scale, Beaufort says, "Or that which will enable a man of war to steer." Of 2, light breeze, he writes, "Or that which will carry a man of war with all sail set 3 or 4 knots." By 4 the same man-of-war is traveling at 6 knots, and by 5, fresh breeze, she must reduce her sails—Beaufort lists the ones she can still carry: "Or that to which whole topsail, topgallants, royals, flying jib, & staysail may be carried full & by." By a wind of force 7, moderate gale, he's describing "that to which the same ship would carry 2ble. Reefed topsail & Jib." That is, the topgallant and royal sails—the two highest on each mast—are down, and the wind is strong enough that the ship is reduced to the two lowest sails, and the topsail is reefed twice. You progressed through triple-reefed sails until, by 10, whole gale, "a man of war could show no other canvas than storm staysails"—the triangular sails that run between the masts, used only to provide stability during the worst blows. At 11, storm, the wind is "that which would blow away any sail made in the usual way." Of 12 he simply repeats himself: "Hurricane!"

And thus, in the Rio de la Plata, a new standard of seafaring description was created.

Beaufort had done something remarkable—he had taken his list of words and, by attaching them to something real, something actual and observable, had suddenly made them meaningful, useful, a self-evident scale that every sailor could be expected to easily apply. The frigate—the three-masted man-of-war—was as close to a standard ship as existed in the early nineteenth century. Even on a ship with different rigging—a bark, perhaps, whose third mast would be rigged differently, or a brig, which had only two masts—any but the newest sailor would know by experience what canvas he'd be showing on a frigate. Every sailor could instantly apply the new scale, called simply "Wind Scale" in Beaufort's journal.

It was, perhaps, a moment of transcendence.

Then again, it turns out, that probably wasn't Beaufort's idea either.

I FIRST REALIZED THERE WAS GOING TO BE more to the Beaufort Scale than some genius of prose sitting quietly, saying, "Aha!" and then writing "calm: smoke rises vertically," when, despite what I had learned in *Beaufort of the Admiralty,* my interest in the scale refused to wane. I spent a long period writing letters and calling libraries, chasing trails I found in footnotes and indexes. I'm not familiar with this kind of research, so my efforts were sort of scattershot, and when I found myself working now and then for a newspaper where I could bury overseas calls in poorly vetted expense reports, I'd make a phone call or two as well. As research, it was pretty half-assed.

I'd send a vague letter of inquiry to, say, the Royal Society, and a few weeks later I'd get a note back suggesting that maybe I try the British Museum. I'd cogitate on that awhile and then six months later I'd send a letter, and a month or so thereafter I'd get a note urging me to contact, say, the Met Office. I was trying to live a productive life at the time, so this wasn't a constant effort. In speed it made the pursuit of that white Bronco with O. J. Simpson inside look like the Daytona 500.

Still, I never quite gave up, either. I kept coming back to the scale in the dictionary, and its prose continued to hypnotize: "small trees in leaf begin to sway," I read again and again in "force 5." And I read it often enough that one day I realized—that's *iambic.* Small *trees* in *leaf* be-*gin* to *sway.* It sounds like Shakespeare because that's how folks speak in Shakespeare. Followed immediately by "crested wavelets form on inland waters," which was trochaic, and not only that but trochaic pentameter. I hadn't been kidding myself when I considered the scale poetic—it was actually poetry. But I still was having trouble getting the story of where the thing actually came from.

I had read the biography of Beaufort, and though I'd read all about January 1806 and Beaufort's journal and so forth, I hadn't read a word about the scale I'd come to love. I had read on and learned a bit about the scale's development; it was pretty cool, and I'll tell you all about it in a little bit.

But nothing about smoke, about leaves, about the kind of vital, sensory writing I was seeking. Finally, though, I did send a brief letter to the Met Office, which led someone there to send me a photocopy of a document referred to as Official No. 180, published in 1906, its title hardly a model of Beaufortian brevity: "The Beaufort Scale of Wind-Force. Report of the Director of the Meteorological Office upon an Inquiry into the Relation between the Estimates of Wind-Force According to Admiral Beaufort's Scale and the Velocities Recorded by Anemometers Belonging to the Office." Translated, that means the people at the Met Office were trying to figure out actual wind velocities to correspond with the different Beaufort numbers. The report is fifty-four pages long, including "A report upon Certain Points in Connection with the Inquiry" by an engineer named Sir George Simpson and notes by two scientists and a navy commander.

I plowed through the report. An awful lot of stuff about anemometer cups and pressure tubes, about overestimation and underestimation and plotting wind over time and comparing curves. I have to be honest—there aren't a whole lot of people more interested in the Beaufort Scale than I am, and even I was not particularly interested in Official No. 180. I got to the formulae for adopted curve of velocity-equivalents ($P = .003V^2$, if you're wondering). My eyes glazed over. And then I turned a page.

And there it was.

Spread across two pages, in a broad table, was the "Specification of the Beaufort Scale with Probable Equivalents of the Numbers of the Scale." At the left were the twelve numbers, followed by Beaufort's terms and the descriptions he used in his revised scale in 1807 based on the sails a ship could carry, with 1 being "Just sufficient to give steerage way" and 12 topping out at "that which no canvas could withstand." There followed a column describing the "Mode of Estimating aboard a Sailing Vessel," such as "Reduction of sail necessary even with leading wind."

14 | 15

TABLE II.
Specification of the Beaufort Scale with Probable Equivalents of the Numbers of the Scale.

Beaufort Number.	Specification of Beaufort Scale. Admiral Beaufort's General Description of Wind.	Admiral Beaufort's 1805.		Description of Wind.	Mode of Estimating aboard Sailing Vessels.	Specification of Beaufort Scale. For Coast Use, based on Observations made at Scilly, Yarmouth and Holyhead.	For Use on Land, based on Observations made at Land Stations.	Mean wind force in lbs. per square foot at standard density. P=·0105B².	Equivalent hourly velocity for expressing individual estimates in miles per hour, and vice versâ (V=1·87√B³).	Equivalent mean velocity for expressing the average of a large group of estimates in miles per hour, and vice versâ.	Probable maximum velocity attained by wind.	Probable mean velocity of wind in gusts.	Probable mean velocity of wind in lulls.	Probable minimum velocity attained by wind.
0	Calm	Calm		—	—	Calm	Calm; smoke rises vertically.	0	0	3	1·5	·5	0	0
1	Light air ...	Just sufficient to give steerage way.				Fishing smack* just has steerage way.	Direction of wind shown by smoke drift, but not by wind vanes.	·01	2	5	4	3	1	·5
2	Slight breeze...	That in which a well-conditioned man-of-war, with all sail set and "clean full" would go in smooth water from	1 to 2 knots	Light breeze	Sufficient wind for working ship.	Wind fills the sails of smacks, which then move at about 1–2 miles per hour.	Wind felt on face; leaves rustle; ordinary vane moved by wind.	·08	5	8	9·5	7·5	4	3
3	Gentle breeze		3 to 4 knots			Smacks begin to careen, and travel about 3–4 miles per hour.	Leaves and small twigs in constant motion; wind extends light flag.	·28	10	11	15	13	7·5	6
4	Moderate breeze		5 to 6 knots			Good working breeze; smacks carry all canvas, with good list. White crest on waves.	Raises dust and loose paper; small branches are moved.	·67	15	15	24	20·5	12·5	10
5	Fresh breeze ...	That to which she could just carry in chase, "full and by"	Royals, &c. ...	Moderate breeze.	Forces most advantageous for sailing with leading wind and all sail drawing.	Smacks shorten sail	Small trees in leaf begin to sway; wavelets form on inland waters.	1·31	21	19·5	30	27	16·5	13·5
6	Strong breeze		Single-reefed topsails or top-gallant sails.			Smacks have double reef in main sail. Care required when fishing.	Large branches in motion; whistling heard in telegraph wires; umbrellas used with difficulty.	2·3	27	21·5	38	34	21	17·5
7	Moderate gale		Double-reefed topsails, jib, &c.	Strong wind	Reduction of sail necessary even with leading wind.	Smacks remain in harbour, and those at sea lie to.	Whole trees in motion; inconvenience felt when walking against wind; umbrellas discarded in exposed places.	3·6	35	30	46·5	42	26	21·5
8	Fresh gale ...		Triple-reefed topsails, &c.			All smacks make for harbour, if near.	Breaks twigs off trees; generally impedes progress.	5·4	42	36	56	51	31	26·5
9	Strong gale ...		Close-reefed topsails and courses.	Gale forces...	Considerable reduction of sail necessary even with wind quartering.	—	Slight structural damage occurs (chimney pots and slates removed).	7·7	50	44	66	60	37·5	31·5
10	Whole gale ...	That which she could scarcely bear with close-reefed main topsail and reefed foresail.				—	Seldom experienced inland; trees uprooted; considerable structural damage occurs.	10·5	59	53	78	71	44·5	37·5
11	Storm	That which would reduce her to storm stay-sails.		Storm forces	Close reefed sail running, or hove to under storm sail.	—	Very rarely experienced; accompanied by widespread damage.	14·0	68	—	—	—	—	—
12	Hurricane ...	That which no canvas could withstand.		Hurricane ...	No sail can stand even when running.	—	—	Above 17·0	Above 75	—	—	—	—	—

* The fishing smack in this column may be taken as representing a trawler of average type and trim. For larger or smaller boats and for special circumstances allowance must be made.

Table II, from Met Office Publication No. 180, published in 1906.
The birth of the modern Beaufort Scale.

But after that, hidden in the middle, came two purely descriptive columns: one "For Coast Use," which is a tiny version of Beaufort's sail-description scale, going from 1, "Fishing smack just has steerage way," up only to 8, "All smacks make for harbour, if near." And then, "For Use on Land, based on Observations made at Land Stations," came the scale I knew: "Calm; Smoke rises vertically," the bits about chimney pots and slates, "umbrellas used with difficulty," and all the rest.

It turns out that the 1906 document was supposed to be the final word in a debate that had percolated for about as long as the scale had been in use. The scale was supposed to determine wind *force* as opposed to wind *speed,* and since by the beginning of the twentieth century there were useful anemometers, this 1906 document was supposed to correlate measured speeds to the different forces of the Beaufort Scale. Sir George Simpson had been in charge of a committee trying to do that, to figure out how fast the wind was going when it was at a particular strength. That is, when the wind was *strong* enough to move your ship at a speed of four knots, how *fast* was it going? To translate the speed of ships—on which anemometers still didn't function well—to wind speed the Met Office determined to place people familiar with ships in observation points at the coast, where they'd feel the wind and watch the ships at sea; then they'd make a Beaufort Scale estimate, and they'd also keep track of what they saw around them and take anemometer readings. The observers were placed in five stations around the United Kingdom, and the results of their observations had gone into the scale that I had found. By all appearances, Sir George was the one who had stitched their observations into the scale I had read. So all the descriptions—all those lovely passages in the dictionary—all that stuff was written in 1906.

By an engineer.

Still, you follow where the trail goes. So, thinking I had found my prose genius, I started looking up writing by Sir George. The rest of his portion of Official No. 180 was disappointingly flat ("The expression 'mean velocity of the gusts' is a very loose term, and cannot be accurately defined; but experience has shown that with a little practice it is possible to obtain a high degree of consistency in measuring such a quantity"), so I went looking elsewhere.

I found a couple of articles he wrote in a journal called *The Nineteenth Century and After,* one about climatic change and another about the history of weather forecasting. They're interesting enough, and he can be rather coy ("From the evidence Stein and Huntington are convinced that there was more water in the rivers formerly, while Sven Hedin considers that all the changes can be accounted for by bad government and drifting sand"), but it's mostly pretty straightforward stuff: "It appears reasonable to associate these changes with those which caused the level of the Caspian Sea to vary and the rivers in the Tarim Basin to alter their length." Plain good prose, to be sure, but trochaic pentameter it's not. But the trail seemed cold: Sir George had either written or, even more distressing, assembled the table. My lodestar of prose had been written at best by an engineer—more likely, by a bunch of them.

One hundred years after Francis Beaufort sat down aboard the *Woolwich* with his pen and ink and scrawled a list of words in his journal, the passages that I thought were the best descriptive writing I'd ever read had been written by committee.

JUST THE SAME, THAT PUBLICATION, OFFICIAL NO. 180, finally gave me a historical sense of the Beaufort Scale, and pieces began to fall into place. For one thing, it brought to the fore something from Beaufort's journal that I had been unable to process. That original scale, the one he wrote in 1806, was, first, not descriptive in the least; second, of no particular use; and third, not even his own work—not his own work one bit.

The story goes on, but let's start with numbers. If assigning numbers to wind was a big development, Beaufort was more than 200 years late.

Start without even leaving England. In 1662 the Royal Society was founded, as England's first organized gathering of men pursuing scientific understanding. It was filled with people like Robert Hooke and Isaac Newton, and its *Philosophical Transactions* were destined to be the organ of publication of scientific advancement for the next several centuries. But most important to this story, one of the very first things the Royal Society did was show interest in the new science of meteorology. In one of the first editions of the *Philosophical Transactions,* Robert Hooke in 1663 described "A Method for making a History of the Weather." And the very first element Hooke urged his readers to observe was "the Strength and Quarter of the Winds." His scheme for representing those observations conveyed wind strength in the numbers 1 through 4, presumably including 0 if the day was completely still. The table Hooke shows as an example includes half-numbers, so this would have been a nine-point scale, though there's no record of any description of what the numbers meant. Still—if people thought of ranking wind from 0 to 4 by half-numbers 140 years before, ranking the wind from 0 to 12 by whole numbers hardly seems like that much of an accomplishment.

As it happens, even Hooke wasn't first to the party; in 1582 no less a luminary than the great astronomer Tycho Brahe, whose consistent observation of planets and stars and his improvement in instrumentation helped astronomy move into its modern era, got interested in the weather. Tycho was the last great astronomer not to be convinced of the Copernican system, whereby the Earth orbits the sun. (He was still more comfortable with the Earth at the center of things, though it called for some pretty fancy theoretical footwork to explain, for example, the planets' retrograde motion; he eventually decided the planets orbited the sun, but the sun orbited the Earth. Wrong, of course, if still rather elegant.) Sill, it was Tycho's meticulous observations, eventually used by his successor, Johannes Kepler, that helped prove Copernicus right. And regardless of whether the Earth was at the center of the universe, Tycho was interested in what happened on it—for example, in how weather varied over time, and in how you could predict it. So he organized a series of meteorological observations on the island of Ven, between Sweden and Denmark. He devised a scale for wind records that started at 0, for dead calm, then had two categories of light wind (*l,* with a comma, and *l* without one, possibly for the Danish *lidet,* little; the comma indicated lightness of degree). Following came five different degrees of *g* (for the old Danish *graa,* which may have meant "windy"), with a following comma or period demonstrating degrees of lightness and a following apostrophe or two indicating harshness. At the high end of the scale were three degrees of *s,* for storm: *s, s',* and *s''*.

Each of Tycho's levels had several accompanying descriptions: *l* was anything from quiet weather to rather quiet, soft, or weak. For *g* he meant, among other things, strong blowing, rather strong wind, rather hard wind, big wind, stiff wind, and so forth. For *g* he is even more inclusive, stating that it represents "indefinite degree of strength, which I have used where words such as 'blowing' or 'windy' are, so to speak, tossed off to indicate the direction of the wind, where they do not seem to indicate anything other than what in our everyday life seems like the common concept of: wind." So Tycho was getting started, but clarity in wind scales had a long way to go.

Tycho continued those observations until 1597, when the arguments about patronage and so forth that beset Renaissance scientists caused him to relocate to Vienna. In the glare of his astronomical contribution, Tycho's meteorological interests have been overlooked, and his *Meteorological Daybook,* as it was called, has been virtually forgotten. In any case, not much came of his fifteen years of observations. Still, his combination of letters followed by symbols was a good idea, and it would show up again.

Hooke's improved suggestion seventy years later to the Royal Society met with a similar fate—his meteorological observation table was sufficiently interesting to be included in the history of the Royal Society written in 1667, but if anyone besides Hooke actually kept any observations, they're not easy to find today. Nonetheless, a good idea is hard to crush, so in 1723, James Jurin, then secretary of the Royal Society, made almost exactly the same request for consistent meteorological observation—in Latin, this time—in the *Philosophical Transactions.* Jurin's wind scale at least included simple explanations: 1 was meant to signify "the gentlest motion of the wind, which scarcely shakes the leaves on the trees"; 4 was the most violent wind; 2 and 3 were in the middle; and 0 was "perfect calm."

Jurin sent copies of his idea to people all over the world (Cotton Mather got one; Leeuwenhoek was another regular correspondent, but he happened to die just then). But like Hooke's, Jurin's request failed to generate much international weather recording. In addition, such recording as it did generate (some appeared in subsequent years in the *Philosophical Transactions*) offers little help if you're trying to puzzle out what the wind was like. Day after day of "2" or "3" yields a classification so broad as to be scarcely helpful—"somewhere between a light breeze and the end of the world" doesn't communicate much.

Still, when you're a hammer, everything is a nail; when you're looking for wind scales, they turn out to be everywhere. In subsequent years wind scales were . . . you could almost say in the wind. In the early 1700s a Dutch fellow named Jan Noppen worked one out that had some seventeen gradations of wind, though that was a lot more than most, and its very complexity may have killed it; if four levels of wind yields too little information, seventeen may just be too much trouble. In any case, no record remains of the descriptions of his scale, if there ever were any.

The Palatine Society, a European meteorological group based in Mannheim, Germany—it was probably the first organized meteorological society in the world—undertook in 1780 to organize standardized weather observations among several cities, and it came up with a five-point scale: 0 was calm; in 1 only the leaves moved on the trees; in 2 small boughs moved, and in 3 large ones—in 4, boughs were torn off. The scale was most likely based on Jurin's scale, though another scientist had created a wind scale in the mid-1700s. His was based on the "great oak" in the garden of his observatory in Uppsala, Sweden, and followed the same five-level gradation, adding for grade 2 that the wind moved a heavy weathervane, and that in grade 4 "the trunk itself swayed vehemently." If his name didn't get attached to his wind scale (which itself, of course, might very likely have been based on Jurin's), don't feel too bad for him, since he came up with another numeric gradation that seems to have lasted. His name was Anders Celsius.

The Palatine Society lasted a decade or so, and then it, like all the other attempts to begin organizing meteorological data, fell apart. Part of the problem was that by the end of the eighteenth century Europe was tearing itself to pieces at war; part of the problem was that, at war or at peace, communication was troublesome and uncertain.

AND THAT'S JUST NUMBERS. People had been listing the wind for centuries—even millennia. By the time Tycho Brahe and other people began looking for a way to make a scale of the winds, the names of the winds were already finding a general agreement.

Aristotle might not have understood what the wind was or what caused it, but his *Meteorologica* included a list of the known winds, organized according to the quarters from which they came. The era of ocean travel, of course, increased the need for clear terms about the weather mariners might encounter, and those who had the most experience turned to the task. Captain John Smith—Pocahontas's friend—sailed back and forth several times from England to the New World, so as early as 1626 he wrote a book called *An Accidence, or The Pathway to Experience Necessary for all Young Seamen,* which appeared the next year as *A Sea Grammar.* In it Smith offered a great deal of knowledge of the sea, including, in chapter 10, "Proper tearmes for the Winds."

Like any good scale, his starts from the bottom: "When there is not a breath of wind stirring, it is a calme or a starke calme. A Breze is a wind blowes out of the Sea, and commonly in faire weather beginneth about nine in the morning, and lasteth til neere night; so likewise all the night it is from the shore. . . . A fresh Gale is that doth presently blow after a calme, when the wind beginneth to quicken or blow. A faire Loome Gale is the best to saile in, because the Sea goeth not high, and we beare out all our sailes. A stiffe Gale is so much wind as our top-sailes can endure to beare. . . . It over blowed when we can beare no top-sailes. A flaw of wind is a Gust which is very violent upon a sudden, but quickly endeth. . . . A storme is knowne to every one not to bee much lesse than a tempest, that will blow down houses, and trees up by the roots. . . . A Hericano is so violent in the West Indies, it will continue three, foure, or five weekes, but they have it not past once in five, six, or seven yeeres; but then it is with such extremity that the Sea flies like raine, and the waves so high, they over flow the low grounds by the Sea, in so much, that ships have been driven over tops of high trees there growing, many leagues into the land, and there left."

The similarities to Beaufort's eventual scale—Smith describes several winds in terms of the amount of sail a ship can carry—are plain. And so is the delight of the language. Describing states of the sea,

Smith says, "We say a calme sea, or Becalmed, when it is so smooth the ship moves very little, and the men leap over boord to swim." He continues to a rough sea and an overgrown sea, ending with "the Rut of the sea where it doth dash against any thing." With little manipulation you could organize Smith's work like Beaufort's, into a neat nine-point scale:

0. Calm: Ship doesn't move; men leap overboard to swim.
1. Breeze: Light movement of air, offshore during the day and onshore at night.
2. Fresh gale: Calm is over; ship can sail.
3. Faire loome gale: Sea not yet rough, but ship travels well with all sails set.
4. Stiff gale: Topsails just able to bear wind.
5. Over blown: Topsails must be struck.
6. Storm: Just below a tempest.
7. Tempest: Houses blown down; trees torn up by the roots.
8. Hurricane: Sea flies like rain and floods low-lying coastal areas.

It's actually quite lovely, and it was written 180 years before Beaufort put pen to paper—and thirty years after Tycho lifted pen from his.

It was only the beginning. In Defoe's *The Storm,* written another eighty years after Smith's work, terms have been even more thoroughly categorized. Before thoroughly describing the terrible storm, Defoe first discusses whether England faces worse weather than the rest of Europe, and he decides that it must, which makes the English sailor such a doughty fellow when compared with his neighbor to the south—don't forget, Defoe wrote *The Storm* about a century after the British had defeated the Spanish Armada, largely because the Spaniards couldn't handle the weather. Defoe doesn't pull his

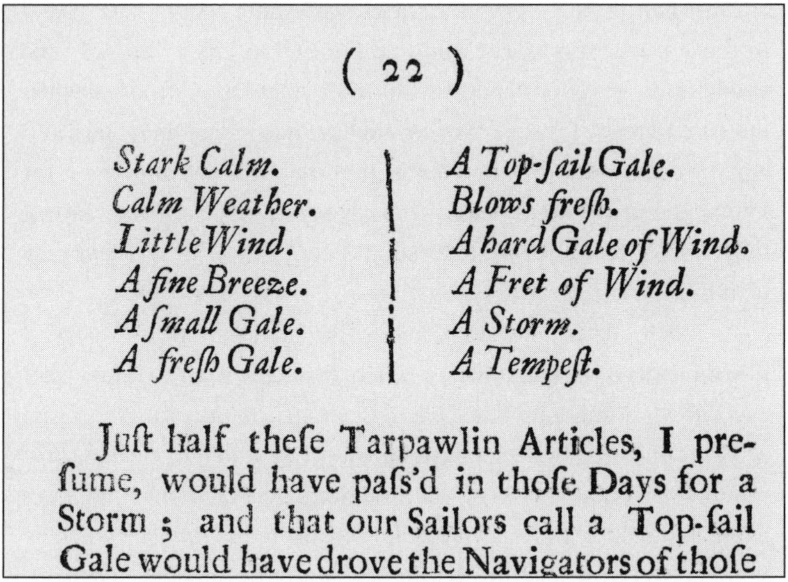

(22)

Stark Calm.
Calm Weather.
Little Wind.
A fine Breeze.
A ſmall Gale.
A freſh Gale.
A Top-ſail Gale.
Blows freſh.
A hard Gale of Wind.
A Fret of Wind.
A Storm.
A Tempeſt.

Juſt half theſe Tarpawlin Articles, I preſume, would have paſs'd in thoſe Days for a Storm ; and that our Sailors call a Top-ſail Gale would have drove the Navigators of thoſe

The list of "tarpawlin articles" included in *The Storm* by Daniel DeFoe in 1704 shows how early terms for the wind were being organized.
From *The Storm* by Daniel DeFoe, p. 22. Special Collections Library, University of Michigan.

punches. He mentions "all the dismal things the Ancients told us of Britain, and her terrible shores," and concludes that "such winds as in those Days wou'd have pass'd for Storms, are called only a *Fresh-gale,* or *Blowing hard.* If it blows enough to fright a South Country Sailor, we laugh at it: and if our Sailors bald Terms were set down in a Table of Degrees, it will explain what we mean.

"Stark Calm.
Calm Weather
Little Wind
A fine Breeze
A small Gale
A fresh Gale
A Top-sail Gale
Blows fresh
A hard Gale of Wind
A Fret of Wind
A Storm
A Tempest

"Just half these Tarpawlin Articles, I presume, would have pass'd in those Days for a Storm; and that our Sailors call a Top-sail Gale would have drove the Navigators of those Ages into Harbours: when our Sailors reif a Top-sail, they would have handed all their Sails; and when we go under a main Course, they would have run *afore it* for Life to the next Port they could make; when our *Hard Gale* blows, they would have cried a Tempest; and about the *Fret of Wind* they would be all at their Prayers."

As much fun as it always is to hear the English make fun of people of other nations, that list especially made me want to know what a fret of wind might be. The *Oxford English Dictionary* says this sense of *fret* is obsolete, but it's a gust (of wind), a sudden disturbance (of weather) or agitation (of waves), and among its citations it includes Samuel Johnson, who used it metaphorically in the *Rambler* when he said he was "frustrated of [his] hopes by a fret of dotage." The concept of a gust of dotage is so much like an eighteenth-century anticipation of a "senior moment" that it makes me wish *fret* in this sense would return to common usage.

Speaking of Johnson, he of course included the standard words for wind in his dictionary of 1755, and under *tempest* he lists them all. A tempest, he says, is "the utmost violence of the wind; the names by which the wind is called according to the gradual encrease of its force seems to be, a breeze; a gale; a gust; a storm; a tempest," and he cites as his source a passage from Donne: "What at first was call'd a gust, the same/hath now a storm's, anon a tempest's name."

OKAY, SO WE KNOW that Beaufort wasn't the first to list the names of the wind in order of increasing magnitude—that's been going on since the 1580s—and we know he wasn't the first to put numbers on a wind scale, which was happening at least by 1663. He wasn't the first to use observation and description either—Jurin's scale had that, as did Tycho's scale, to a degree. But there's that wonderful moment in 1807 when Beaufort decided to organize the winds according to the sails a ship could carry. A moment of genius, yes?

Sure, but probably not Beaufort's genius. It was a great idea, for example, when the Swedes were doing it in the late 1700s—a Swedish scale was written down in 1779, when the crew of a sixty-gun Swedish ship used a handheld deflection anemometer to take wind pressure readings, at which point the officers would confer and identify the wind. This scale sensibly skips the low breezes, at which little was going on to interest a sailor, and starts from a "lab. Bramsegels Kultje," which can be translated to mean "light topgallant breeze," and progresses through topgallant breeze and fresh topgallant breeze up through a series of topsail winds—they would have been called "topsail gales" in England—including Stiff Topsail Wind, Reefed Topsail Wind, and Stiff Reefed Topsail Wind. (Tage Andersson of the Swedish Meteorological Society, who brought these to my attention, suggests they could more accurately be thought of as single-, double-, and triple-reefed topsail winds, which makes them even more directly ancestral to Beaufort's scale.)

This was happening all over. A Dutch scale developed in the seventeenth century has as its lowest category "a dying breath" of wind, moving up through a "listless wind" before coming to terms like "topgallant sails wind," "topsails wind," and "course wind," all describing the sail of the ship—on the way you also ran into winds like a "muzzler," which was blowing against you, or a "rogue wind," which brought a storm.

I NEVER KNEW SO MUCH about wind, and its names, and the order of those names. A fret of wind? Defoe thumbing his nose about ancient sailors from the south? John Smith explaining that he could tell when the sea was calm, because the sailors started leaping off the ship—perhaps holding their noses and wearing inflatable seahorses around their midriffs? That's fun stuff.

And it was part of what drew me to the Beaufort Scale. The beautiful language, I could see, didn't appear suddenly in the captain's cabin of the *Woolwich,* whether offshore of England while Beaufort was awaiting orders, or in the Rio de la Plata the night before he sailed for home. The order and language of the scale had predecessors, and plenty of them. There was something about describing the wind that sparked expressive language.

I got to wondering what that was.

And I think the answer is that the wind is invisible. You can't describe it because you can't see it. You can only describe what it does to things that you *can* see—sails, the sea, trees, roof tiles. To describe clouds, trees, or anything else, you focus in on that specific thing, ignoring everything else. To describe the wind, you do the opposite: you look at everything else. It's mind-expanding. But since not everybody who observes the wind is great at writing clear description, a scale solves a real problem. In *The Storm,* written after the great storm of 1703, Defoe himself made much of the fact that nobody knew why the wind blew. In only thirty years Hadley cleared the matter up, which is great. But one of the contributions to Defoe's book pointed out another deficiency—the lack of any way actually to measure a storm.

William Derham, a member of the Royal Society, wrote a piece for the *Philosophical Transactions* about the storm that Defoe included in his book, and Derham makes the point straightforwardly: "The degrees of the Wind's Strength being not measurable (that I know of, though talked of) but by guess," he says, he can think of measuring the high wind only "with respect to other storms.

"Feb. 7 1699 was a terrible storm that did much damage—this I number ten degrees," he says. On February 3, 1702, on the other hand, was the "greatest descent of mercury every known: This I number 9 degrees. But this last of November, I number at least 15 degrees." That is, this was the *Spinal Tap* of storms—this one goes to *fifteen.*

Derham's comments make clear that nobody had yet devised a useful anemometer ("though talked of"). More important, they show that despite the different wind scales and lists floating around, certainly none was widely known. But it was becoming clearer that one of them had to be—if people were going to share information about the wind, they were going to have to know what they were talking about. Derham recognized that the guys sitting around the docks saying, "You thought that was wind? Compared to back in '99, that was *nothing!*" were at a dead end; without any kind of comparative scale, any disagreement about storm or wind intensity would have to end up at he-said-she-said. Derham's little attempt at classification shows that it was more than just mariners who needed to be able to describe the wind. That attempt brought to mind one of the subtexts that make the Beaufort Scale so attractive.

These words do *work.* These words have a *job*—to make you understand how the wind is blowing, *exactly* how the wind is blowing, in comparison with other winds. They express perfectly a fundamental thing about language:

Language is technology. It's a tool to accomplish a task. I sat in on part of a course about the history of science and technology, and the professor began with a delightfully pithy timeline that he called the history of technology in twenty minutes. Science, he pointed out, was systematic knowledge about the real world; technology, on the other hand, comprised things we create to solve problems, to do work.

And the first piece of technology he cited, created millions of years ago when a stone tool was—quite literally—the cutting edge, was language. Language is, at bottom, a tool, a technology for communication. And irrespective of its beauty, the Beaufort Scale, in its clarity and specificity, is a spectacularly sharp tool.

Like the sailing ship. For people in the eighteenth and nineteenth centuries, wind was more than a phenomenon. It wasn't just weather—in fact, the term "wind and weather," which has been around since Aristotle, points out that the two were not considered the same thing. As I had discovered on the *Europa,* wind was a way to do work.

A PRETTY ACCURATE WAY to peg the beginning of the Industrial Revolution is verbally. Everybody knows James Watt patented an improvement to the steam engine in 1769 that solved many of the problems of that new source of power, rendering it finally useful. It may be equally important, though, that Watt also invented a word: horsepower.

That's important because it points out that until then nobody needed to talk about horsepower because horses were about the only kind of ground power there was—there was nothing to compare horsepower *to*—nothing to describe in terms of horsepower except horses, each of which has a horsepower of 1.

On the other hand, there was one other source of cheap and useful power before steam and coal. That power was wind. And the two greatest machines of the preindustrial world were wind machines. In the wind scale–rich environment of the late eighteenth century, those two machines—through men intimately familiar with their function—combined to create the Beaufort Scale. One was the sailing ship.

The other was the windmill.

CHAPTER 4

Reverse-Engineering the Wind

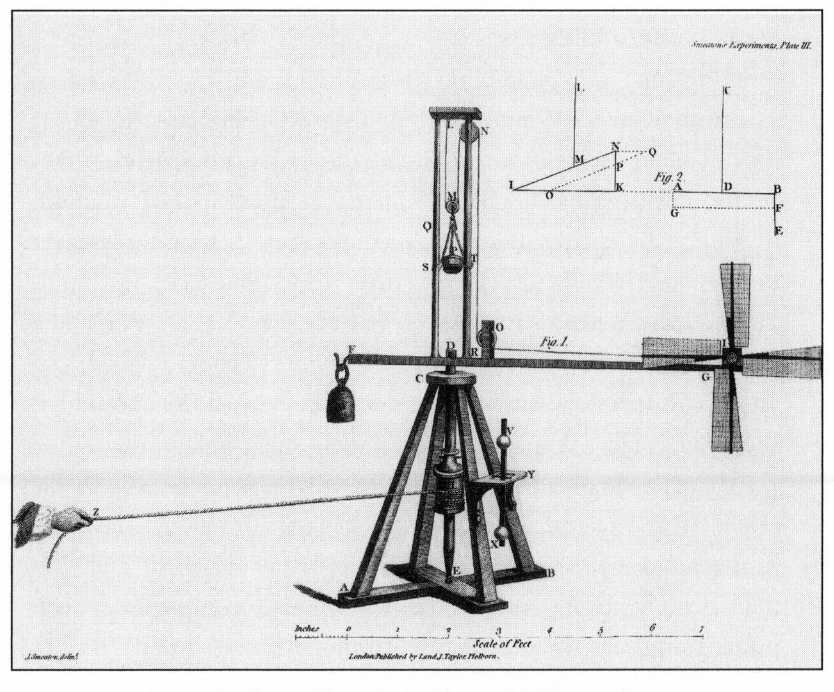

John Smeaton's drawing of the apparatus he used for his wind experiments. The wind itself was unpredictable and he found it easier to move the device through the air than push air through the device.

From *Experimental Enquiry Concerning the Natural Powers of Wind and Water…* by John Smeaton. Special Collections Library, University of Michigan.

EVERY WINDMILL, it turns out, has its own name. Or every Dutch one does, anyway. In 1693, reflecting the importance of the mills that provided power and kept the sea at bay, a Dutch edict required each mill to have a name, which identified it and functioned like an address. So, instead of just going to the local windmill for your flour, you'd go to "Pale Death," to "The Colorful Hen," to "The Stork." You could read the name on what's called the beard, a fancy placard that corresponds roughly to the figurehead on a ship.

In the United States, you can see a beard on *De Zwaan* ("The Swan"), a four-bladed wooden windmill, as tall as a twelve-story building, that now stands on "Windmill Island," in the western Michigan town of Holland, overlooking grassy fields and the obligatory ye olde Dutch village. *De Zwaan,* though, is completely authentic, the only working Dutch windmill in the United States. It was built in the 1720s, and when it came to Michigan in 1964, it was the last ever to leave the Netherlands. (In 1850 there were about 9,000 windmills in the Netherlands; by 1950, as a result of electrification, war, and a cultural blind spot, only about 900 remained. People noticed, and after *De Zwaan* they slammed the door.) In the Windmill Island gift shop you can buy flour, because *De Zwaan* still grinds wheat.

"We're looking for wind of about twenty miles per hour," says the miller, Alisa Crawford, who, when she's not actually milling, shows visitors around the slatted wooden innards of the octagonal mill. Less than 15 mph and it's hard to keep the wheels turning with enough power to manage the grinding. Much more than 25, and it's hard to control the blades and harder to stop them. It takes a good five minutes, she says, to get things slowed to a stop when the wind is rising, and a sudden stop courts disaster. "The analogy is a car," she says. "You don't want to get going too fast and have to slam on the brakes." The windmill uses a drum brake, and that too has limits: braking too fast can cause whiplash, and if one of the 40-foot blades breaks off, it can kill people. Brake too slowly, on the other hand, and the friction of the wooden brake can burn off the beeswax lubrication and set the wooden drum afire—just like those windmills in Defoe's storm of 1703.

Crawford clambers around the inside of the mill, showing off the wooden gears (to prevent wear, the teeth of adjoining gears must be of different kinds of wood), the huge millstones (about a ton apiece; they last virtually forever), and the ingenious governor the mill used to control the speed and the separation of the millstones. Metal balls about the size of those used in the shot-put depend from levers on a wheel; if the wheel goes faster, the miller can allow the increased force to push the balls farther out, slowing the rotation. "That way if the blades slowed down you could separate the millstones and let the drag decrease and blades pick up speed," Crawford explains. "To accommodate wind variability. It's not like a water mill. The water is a lot more predictable." In fact, she says, she believes only 25 percent of working a windmill has to do with actual milling: "I spend 75 percent of my time watching the weather."

The governor underscores the windmill's status as a miraculous machine, and reminds you that it's more than a little like a ship, like the *Europa*—and it looks the same, too. It has the same wooden construction, the same dependence on the wind and on the power of people to be ready for the wind. Crawford points out a crossbeam above the mill's highest floor that she says reminds her of a decorated thwart in a ship; and *De Zwaan* has a vertical capstan that Crawford can work on her own, by walking on it, which turns the entire top of the windmill into the wind; that's almost exactly like the horizontal capstan a tall ship uses for raising anchors and yards. And as with a ship, the worst thing that can happen to *De Zwaan* is to be taken by the wind from the wrong side; it's vitally important that when the wind is rising too high, the windmill is oriented into the wind, its blades chained and locked down.

The windmill even uses sails to cover its blades, and Crawford sets her sails according to the wind. As for how she gauges the wind?

"People ask—what kind of wind are you looking for? So I use this." She shows a sheet she copied from a kite catalog specializing in parafoils. It describes the wind from zero to thirty miles per hour, breaking it into seven categories. "3: Gentle Breeze: Leaves and small twigs in constant motion; wind extends light flag."

It's the Beaufort Scale. An American flag flies from the windmill, and she knows that if it's fully extended the breeze is getting toward moderate, around fifteen miles per hour. "If I come out and the flag is snapping out straight, we've got a little power."

It's no surprise she applies the Beaufort Scale to a windmill. What she doesn't know is that using the Beaufort Scale to measure the force of wind for her windmill elegantly closes a little circle that starts about 260 years ago, with an Englishman named John Smeaton, born in 1724, who made one of the Beaufort Scale's earliest contributions. You might say that if people like Tycho and Celsius were the scale's distant forebears and cousins, Smeaton was its great-grandfather, the first direct ancestor with whom you begin to note family resemblance. Smeaton made, almost offhand, an advancement to wind scales that led to further development by Alexander Dalrymple, which led eventually—and quite directly—to what Beaufort finally wrote in his journals. The story of these contributions was until recently unknown, and the people who uncovered the influence of Smeaton and Dalrymple, especially a modern-day researcher named Andrew Cook, are important to understanding the scale's long odyssey from its origins to its current place of authority in the Merriam-Webster *New Collegiate Dictionary.* The place to begin is with Smeaton.

JOHN SMEATON WAS INTERESTED IN EVERYTHING.

"His playthings," says a popular biography from 1844, "were not the toys of children, but the tools men work with." Like any little kid, he loved to watch construction and the men at it ("and to ask them questions," the biography notes pointedly). But Smeaton was the type of kid who, if he saw men fixing a pump, would ask for an extra length of pipe and with it make a pump of his own, that worked. When someone took him to see how millwrights worked, he went home and built a model windmill himself, then climbed on top of the barn to attach it there.

He was six years old at the time.

He never stopped liking windmills—and anything else he could build or create. By the time he was fourteen he was designing engines to do millwork, and according to his biography he even forged his own iron and steel. His parents tried to force him to become a lawyer, but instead, as a lover of astronomy, he first became an instrument maker. By the time he was thirty his skill had made him a fellow of the Royal Society in London. In 1755, the Eddystone Lighthouse—the replacement of the one washed away during the great storm of 1703—burned down, and Smeaton was chosen to rebuild it. He completed it in 1759; in doing so he developed a technique for dovetailing stonework that is still used today. Smeaton based his tremendously successful design "on the natural figure of the waist or bole of a large spreading *Oak*," he wrote, and it seemed to have that tree's grace and strength. It too was eventually replaced in 1882, but only because of advances in lighting technology. As a landmark Smeaton's lighthouse was so beloved that when the new light was completed, Smeaton's was simply re-erected on the Plymouth shore, where tourists visit it even now. By the time he died in 1792, Smeaton had not only helped define his profession but had actually coined the term *civil engineer.* (It's probably worth pointing out that in the Mansell Publishing *History of Technology* of 1986, in an article on Smeaton's papers, the book notes, "It appears from the letters that Smeaton's office organization may have gone somewhat to pieces after his death." A thousand engineer jokes leap to mind, but how profoundly good must an engineer be for it to be a surprise that his desk doesn't remain neat *even after he's dead?*)

But we were talking about windmills—and his work with windmills is what brings Smeaton into the history of the Beaufort Scale. Windmills were hardly new technology—they appeared in Persia sometime between A.D. 500 and 1000 and had made it to Europe by 1200 or so. But a man interested in engineering at the dawn of the Industrial Age was going to be interested in pumps and mills, and windmills had reached their zenith in the Netherlands, so in 1755 Smeaton took a journey there to tour mills and the pumps used for controlling the seawater that would otherwise flood the country; it's not impossible that among his stops he visited *De Zwaan* in its original setting near Amsterdam. When he returned, apart from his work on the Eddystone Light, he wrote up the results of some experiments he conducted on "the Natural Powers of Wind and Water to turn Mills and Other Machines Depending on a Circular Motion" and read the results before the Royal Society.

Because "the wind itself is too uncertain to answer the purpose" of repeated experiments, Smeaton constructed an ingenious model windmill that spun around the perimeter of a circle. Knowing the diameter of the apparatus, the time the model spun, and the number of circles it made, you could measure the speed of the "wind" thus created and measure its effects on the mill at different settings. Among his conclusions Smeaton determined the best form and position of windmill sails: they should be set, for example, not at 90 degrees, perpendicular to the wind, but between 72 and 75 degrees. Smeaton also derived conclusions about what sail velocity would produce the maximum work of the mill (it should equal the wind velocity) and the ratio of the velocity of the ends of the sails to wind speed (unloaded sails—that is, those not doing work—could go up to four times the speed of the wind). The paper was so well regarded that it took the society's Copley Medal, that day's equivalent of the Nobel Prize.

60 EXPERIMENTAL ENQUIRY, &c.

be obſerved, that the numbers in col. 3. are calculated according to the ſquare of the velocity of the wind, which, in moderate velocities, from what has been before obſerved, will hold very nearly.

TABLE VI. *containing the Velocity and Force of Wind, according to their common Appellations.*

Velocity of the Wind.		Perpendicular force on one foot area in pounds averdupois.	Common appellations of the force of winds.
Miles in one Hour.	Feet in one ſecond.		
1	1,47	,005	Hardly perceptible.
2	2,93	,020	Juſt perceptible.
3	4,40	,044	
4	5,87	,079	Gentle pleaſant wind.
5	7,33	,123	
10	14,67	,492	Pleaſant briſk gale.
15	22,00	1,107	
20	29,34	1,968	Very briſk
25	36,67	3,075	
30	44,01	4,429	High winds.
35	51,34	6,027	
40	58,68	7,873	Very high.
45	66,01	9,963	
50	73,35	12,300	A ſtorm or tempeſt.
60	88,02	17,715	A great ſtorm.
80	117,36	31,490	An hurricane.
100	146,70	49,200	An hurricane that tears up trees, carries buildings before it, &c.
1	2	3	

John Smeaton included this table in his 1759 engineering work about the power of the wind. He was already connecting descriptive wind names with specific velocities and, at the last, even light verbal description.

From *Experimental Enquiry Concerning the Natural Powers of Wind and Water...* by John Smeaton. Special Collections Library, University of Michigan.

But most important, in his paper Smeaton included a table "*containing the Velocity and Force of Wind, according to their common Appellations.*" This table broke the wind into eleven categories (not including calm), from "hardly perceptible" through "Gentle pleasant wind" through "High Winds" and "A great storm," to "An hurricane that tears up trees, carries buildings before it, &c." For each category he includes a wind speed in miles per hour and feet per second—and the perpendicular force on a one-square-foot area ("Hardly perceptible" is 1 mph and 0.005 pounds of force; the highest force, tearing up trees and so forth, is 100 mph and 49.200 pounds per square foot).

Smeaton says the table "was communicated to me by my friend Mr. Rouse, and . . . appears to have been constructed with great care, from a considerable number of facts and experiments." (He notes also that a Mr. Ellicot and a Mr. Robins were working on wind measurements, using models not unlike the one he used.) The same table appears in 1801 in John Capper's *Observations on Winds and Monsoons,* attributed again to "Rous." Samuel Rouse turns out to have been a draper by trade, but a mechanic and astronomer by avocation—much like Daniel Augustus Beaufort, father of Francis, he was more distinguished by his energy and friends than his wealth; he died bankrupt in 1775. But his mention in Smeaton's paper of 1759 apparently brought his family such pride that it was still considered worthy of mention in the obituary of his son, who died in 1823.

(The measurements of force that Smeaton came up with were so widely accepted that early experimenters in flight called the value used to calculate drag the Smeaton Coefficient. It turned out he was slightly off in his calculations, which kept Otto Lilienthal of Germany on the ground, though the tables of lift and drag that Lilienthal created based on Smeaton's numbers were widely circulated. In 1901 two guys turned a bicycle wheel on its side, attached a flat plane on one side and a curved plane on the other, and created force by mounting the wheel on a bicycle, which they rode into the wind. They eventually determined that Smeaton's numbers were off by about a third, and they were right—that is, they were Orville and Wilbur Wright.)

Smeaton was also a corresponding member of the Lunar Society, the exclusive group of Birmingham gentlemen (including such luminaries as Erasmus Darwin, grandfather of Charles; Josiah Wedgwood, potter and canal developer; and Joseph Priestley, who discovered oxygen), who met under the full moon to discuss science and its application to industry, beginning in 1765. One of the group's special concerns was accurate measurement—whether of temperature, for use in metalwork, or of weight, for use in manufacturing and experimentation. Or, of course, of wind speed, for its use in milling and pumping. The society had vanished by the early 1800s, but Smeaton's scale was still kicking, on its way to its place in the Beaufort Scale. In fact, it's lasted on its own, as well—in the New York publication *The Engineering News,* September 14, 1889, an article on an electric railway computes wind pressure on a locomotive according to "Smeaton's scale, the best we have."

But to reach Beaufort, Smeaton's scale needed to travel through one more person.

IN THE SECTION ABOUT THE BEAUFORT SCALE in the 1977 biography *Beaufort of the Admiralty,* a tiny footnote says that "there is convincing evidence that it was Dalrymple who suggested to Beaufort the idea of adapting an earlier wind scale by John Smeaton to maritime needs."

Well, you might say so, though that turns out to be rather an understatement.

Barely mentioning the contributions of Smeaton and Alexander Dalrymple to the Beaufort Scale is actually an odd elision, since an article from 1967 in a publication called *The Mariner's Mirror* (the biography cites it) includes Beaufort's first scale (the one that included only words, not descriptions of sail) and Smeaton's scale side by side. A 1968 follow-up to that article noted that *A New Universal Dictionary of the Marine,* which appeared in 1815, gives under the term "Breeze" the following definition: "Mr. Dalrymple, late Hydrographer . . . scientifically arranged the wind in the following order: (1) Faint Air, (2) Light Air, . . . (12) Storm." Author Alfred Friendly's decision to include only a note hinting at the scale's predecessors was probably the result of admiration for Sir Francis rather than anything else, but still, the evidence was there.

The evidence has since then become inescapable. By 1983, according to another *Mariner's Mirror* article, enterprising researchers had discovered a copy of a pamphlet written by Dalrymple in 1779 that included that very listing of the winds, from "1. Faint Air, i.e. just *not calm*," to "12. Storm." Dalrymple had designed it for use in ships' journals, and he produced the pamphlet in his capacity as hydrographer to the East India Company. That pamphlet can be found on paper in the Library of Congress and on microfilm in countless other libraries. What's more, Dalrymple's will (he died in 1808) specifically mentions "a series of prints of sea pieces to explain the degrees and gradations of wind from a Calm to a Storm . . . for the Treatise on Navigation which I have printed wherein is the list for the said Gradations of Wind taken from the Sea Journals compared with Mr. Smeaton's Scale from the Work of Windmills." Nobody has ever found the series of prints (if it even exists), but in recent years the treatise on navigation has itself surfaced (it's called *Practical Navigation,* and though it was never published, there are two incomplete proof copies, one in Paris and another in Scotland), and it includes a version of Smeaton's scale even more similar to Beaufort's most famous version of his wind scale.

This table appeared in *Practical Navigation* around 1790, but it showed up in Smeaton's papers about a decade previously. It separates the wind into nine categories, from Calm to Storm. Most important, though, it does with a windmill what Beaufort later did with his ship. That is, Smeaton defines "Light Breeze" as "Direction

PRACTICAL NAVIGATION. 41

71. TABLE of COMPARISON of WINDS, from Ship's Journals, with Mr. SMEATON's Scale from Auſthorpe Mill, the Length of the *Sails* being 34 feet from the *Center*; or 68 feet diameter.

My Scale.	Mr. Smeaton's Scale	and His Deſcription.	French Terms.
0 Calm	0 Calm	The *Motion* of the *Air*, not felt . .	0 Calme.
1 Faint-Air, . . . i. e. juſt not quite calm	Scarce a Breeze . . .	D° . . . . ſcarcely felt	1 Petit fraicheur, ou foible.
2 Light-Air	Light breeze not working	The *Direction* of the *Wind*, ſenſible, but inſufficient to move the Mill, or under 6 turns in a *minute*	2 Fraicheur.
3 Light-Breeze	1 Light working Breeze .	Juſt ſufficient to move the Mill 6 turns	3 Petit frais, ou petit briſe.
4 Gentle-Breeze	2 Breeze	Sufficient to move the Branches of Trees, and Mill from 6 to 9 turns . . .	4 Jolie briſe ?
5 Freſh-Breeze 6 Gentle-Gale	3 Freſh Breeze	Move the Boughs with ſome noiſe, Mill 9 to 13 turns	5 Jolie frais ? 6 Vent peu de frais.
7 Moderate-Gale . . .	. . . .	. . . .	7 Vent moyenne frais.
8 Briſk-Gale	4 Freſh	Wind heard againſt ſolid Objects and agitation of Trees, Mill from 13 to 18	8 Vent frais.
9 Freſh-Gale	5 Very freſh	Wind growing noiſy, and conſiderable agitation of Trees, Mill 18 to ¾ Cloth	9 Bon frais.
10 Strong-Gale	6 Hard	Wind troubleſome, larger Trees bend, ¾ to ½ Cloth	10 Grand frais.
11 Hard-Gale	7 Very hard	Wind very loud and troubleſome, large Trees much agitated, Mill ½ Cloth to cloſe ſtruck	11 Vent fort.
12 Storm	8 Storm	Wind exceeding loud, Trees very much agitated and ſome broke, Mill 25 to 30 turns without Cloth	12 Tempete.

72. The 7th Term, in my Scale, is a middle term; and, although

The Rosetta Stone of the Beaufort Scale. This table, included in Dalrymple's never-published *Practical Navigation,* took Smeaton's scale based on the motion of a windmill and applied it to words used in ships' journals to describe the wind, perfectly anticipating Beaufort's eventual decision to go a step further and base the descriptions directly on the sails of a ship.

of wind sensible, but its force not sufficient to move the mill, or under 6 revol." per minute; Breeze comes next: "Sufficient to move the branches of the trees, and to turn the mill sails 6 to 9 revol." per minute. Fresh Breeze "moved the boughs with some noise" and turned the mill up to thirteen times a minute; a Fresh Wind is "heard against solid bodies," and a Very Fresh Wind becoming "noisy, with considerable agitation," and the sails on the mill blades have to be reduced to three-quarters cloth, but even then turn eighteen times a minute. The scale ends at Storm, with the wind "exceeding loud and troublesome, large trees very much agitated and some broken," and the mill spinning full speed with no cloth at all.

Smeaton has done here what Beaufort is famed for doing, taking the ineffable wind and finding a way to describe it with something measurable, something sensible, something common.

It makes perfect sense—Smeaton, the greatest engineer of his time, was merely examining the most fully developed machine of his time and gathering as much information as he could about its function, then organizing that information in the most useful way he could. It was fairly simple: Smeaton was an engineer, he worked with windmills, and he was looking for a way to calibrate them depending on the force of the wind. He found that, and then he moved on. He had solved a small problem people had been working on—the weather observers like Tycho and Jurin, the Dutch and Scandinavian sailors—for some time, but for him it was just a small solution to a small problem. That was it. He hardly considered it one of his great accomplishments, and in any case he was much more interested in solving problems than in attention. Catherine the Great offered honor and riches in an attempt to hire him away to Russia, but he preferred England, where he kept hard at work on canals and quays and harbors until his death in 1792. Once done with it, he apparently didn't think much about his wind scale.

But he obviously mentioned it to Alexander Dalrymple. And that took the scale to the next level.

ALEXANDER DALRYMPLE WAS BORN IN 1737 and grew up, like many in England in those days of expanding trade, fascinated with the Orient. He thus started his career at age fifteen aboard East India Company ships as a writer—a sort of historian, documentarian, and communications officer. Dalrymple found a friend in his first captain, and in his time posted in India he had access to valuable libraries and influential people. He became convinced of the possibilities for expanded trade in the East Indies, whereupon the company sent him on a several-year odyssey of sailing, commonly in uncharted waters, and of research. Early in a 1759 voyage to Canton (via Indonesia and the Philippines) to seek trade routes, Dalrymple recognized that nothing would better improve the company's trade than accurate charts of the new territory he saw. Since nobody else was around, he took up surveying on his own and began making those charts himself.

By the time he reached England again, in 1765, he had become an accomplished surveyor and had vitally useful charts—and an essay on nautical surveying—to publish. At about that time the Royal Society was planning a voyage of circumnavigation, to witness the transit of Venus from Tahiti and then to search for the fabled great southern continent of Terra Australis. The Royal Society chose Dalrymple as commander of the expedition and assigned him to a ship provided by the Admiralty. Concerned about his ability to do his work freely with his expedition under someone else's ultimate command, Dalrymple demanded that he command not just the scientific work but the ship itself. But it was the Admiralty's ship, and they refused to acquiesce, so the Royal Society relinquished the command of the HMS *Endeavour* to the Admiralty's choice—a man named James Cook. And so it is that Cook's three voyages of discovery in the uncharted waters of the Southern Hemisphere (he circumnavigated New Zealand, explored the polar region and the eastern half of Australia, and was the first European to see Hawaii, among other things) are the ones we read about today, not Dalrymple's. Dalrymple stuck with the East India Company, becoming in 1779 its first-ever official hydrographer, and in 1795 he likewise became the first Hydrographer to the Admiralty. Just the same, he was never particularly cheerful about Cook.

WITH THE EAST INDIA COMPANY, Dalrymple's hydrographic efforts had a very clear goal: to improve the fortunes of his employer.

Thus in 1779 Dalrymple published a pamphlet for distribution to East India Company ships, suggesting a specific form for the log books all captains were required to keep. Employed in what he called "the very useful work of examining the Journals of [East India Company] Ships, for improving the Charts in the Navigation of the East Indies," he suggests in this pamphlet some specific data that would help him get the clearest understanding of what the ships experienced at sea. The point was simple: "to determine with great precision What is, at every Season, the most eligible course to pursue out and home."

That's all he wanted—Dalrymple wasn't thinking about science or observation, and he wasn't even, at least immediately, all that interested in the increase of general knowledge. He wanted to know how to get his company's ships back and forth to the Indies faster and more safely than the ships of the other guys, and he wanted to be able to figure it out as easily as possible. If ships regularly found helpful or troublesome winds in particular areas, if they regularly encountered storms in certain spots, then having this data would enable the main office to do a better job of plotting future trips. And the more specific the information, the better the conclusions. In that service, to identify prevailing wind and ocean currents, he suggests adding, among other things, a column in the log book for the force of the wind. That's where he included his numbered list of winds, predating Beaufort's by close to thirty years.

What's more, as a fellow of the Royal Society, Dalrymple met Smeaton, and the two became friends. So when Dalrymple got to work on his *Practical Navigation*, poring over ships' logs and finding the terms sailors used for wind vague at best, he turned to the scale designed by his friend. "Mr. Smeaton's is not an *arbitrary* or *fanciful Scale,* but formed by proportionate Work done by the Mill," Dalrymple wrote in *Practical Navigation,* "and it was satisfactory to find, that the *Sea-Terms,* in general, very readily compared with his Scale." And so Dalrymple included Smeaton's scale in *Practical Navigation,* written around 1790.

Beaufort, though, probably never saw *Practical Navigation;* it was never published, and all that remains of it now are those two proof copies. One is buried in the papers of Dalrymple's father in the National Library of Scotland, the other in the Archives Nationales in Paris. It's quite certain, though, that Beaufort saw the 1779 pamphlet including Dalrymple's original list. I know this because Andrew Cook told me so.

AS MY INTEREST IN BEAUFORT AND THE SCALE that bears his name expanded, I found myself corresponding with and visiting researchers and libraries, which was tremendously exciting. As an undergraduate in college, I definitely envisioned my future as a sort of *New Yorker* cartoon of an academic, flitting into and out of the Royal Society and the Library of Congress and such, stroking my chin thoughtfully.

After a few library trips, even though I was pretty lost in most of them, I thought that was me. Beaufort's letters and journals reside at the Huntington Library in Pasadena, for example. When they finally gave me the okay to come and paw through them, I thought I was something—until I arrived at the library and went through an orientation process so bewildering that it took me most of the week I spent there to get my bearings and actually find anything like what I was looking for.

My favorite thing about the Huntington was that fifteen minutes before the end of the day, when you have to turn in your rare books and manuscripts for safekeeping, they tap a little brass counter bell—twice, exactly, rapidly—and you must bring the material back to the counter. At the Clements Library at the University of Michigan, for the same purpose they tingaling a little bell like the one the old rich lady in a stage farce would use to summon her French maid. At the Met Office, they simply come around and sadly tell you they're terribly sorry, but you just have to leave, though you're so welcome to come back again tomorrow.

This isn't the only way libraries differ. At the Huntington, for example, when you are in the sanctum sanctorum of the manuscript room and you whisper a question across the counter, the person there is so likely to know the exact answer that you begin to take it for granted. At the Met Office, the archivists would delight in helping to figure out what I was trying to find, and would invite me to join them for afternoon tea. They did that at the Admiralty Archive, too, but there a friendly archivist even drove me to the train station after I had kept him late by an hour. Perhaps he just wanted to be sure he was rid of me.

At the Library of Congress and the British Library, on the other hand, you are dealing with government employees, and you know it. Both of those libraries are closed-stack, which means the surprising juxtapositions and miraculous leaps forward that come from browsing among the shelves are forbidden: you write the book you want on a slip of paper and someone supposedly goes to get it for you. In the Library of Congress your book comes up on a conveyer belt; at the British Library someone just brings it up—if it ever comes. If it doesn't, nobody can tell you why.

In the British Library restaurant I grumbled about this to other researchers, and they told me horror stories. "They say the British Library is the most complete library in the world," one said. "They keep it that way by not showing anybody the books," and we tittered self-righteously. You have to get a special reader's card to use the British Library, and with my reader's card and my in-jokes I felt like that effete researcher I had always imagined myself one day being.

Then I ran into Andrew Cook—or, as his e-mail signature lists him, Andrew Cook, MA, PhD, FRSA, FRHistS, Map Archivist, India Office Records, The British Library.

"I DO HAVE A LEVEL OF USELESS INFORMATION which I frankly think is emulated by few others," Cook told me with a smile when I finally met him at the British Library. He looks like the Madame Tussaud's waxwork of the Friendly Professor—a tad portly, a full, bushy, graying beard, and a seemingly limitless store of facts and answers. He told me that the research supervisor of his PhD had called him "ridiculously erudite." For his erudition I can vouch; whether it's ridiculous is a matter of opinion, though to me it appeared only extremely helpful. But when I found Cook and began asking him questions, I immediately stopped thinking of myself as anything other than a naïf among records and books. I had found Cook's name attached to a reference somewhere, and I found his e-mail address and began sending him the occasional query. If I had a question about Smeaton, Dalrymple, and Beaufort, Cook could almost always answer it, and almost always immediately. He was unfailingly generous with his time, but his tone occasionally betrayed his frustration with the level at which I was working. He at one point suggested that I check for a letter between Dalrymple and Beaufort, which during my time at the Huntington Library I had not run across. "For God's sake, man," he said, "there *is* a table of *contents*."

There is? No matter. In any case, in Cook I had found, at least for the Beaufort Scale's passage through Alexander Dalrymple, the final source (he wrote his PhD on Dalrymple's manuscripts). For example, I also found a copy of Smeaton's final wind scale—the one where he categorizes the levels of the wind according to their effect on the windmill—in some engineering comments written in the mid-nineteenth century by the final authority on Smeaton's manuscripts. Those comments seemed to indicate that Smeaton and Dalrymple together compared their scales in 1784, and I contacted Cook about it.

"I've looked at the Royal Society Dining Club" records, Cook told me. "I don't think I could find a dinner where the two of them coincided." He also had traced the two existing copies of Dalrymple's *Practical Navigation,* so when I wondered, given the similarity of the wind scale there to the one that has eventually borne Beaufort's name, whether Beaufort hadn't copied that one in 1806, not the mere list in Dalrymple's pamphlet, Cook did everything but snort.

"You've got to posit he saw one of those [two manuscript copies]," he said. "Now, the Paris one is in the former collection of the Societé de Geographie," which was then struggling through the rough seas of revolutionary France. "You'll be aware that there were some political upheavals in Paris in the 1790s?" Cook said archly to me. In any case, he felt sure Beaufort wouldn't have had access to the Paris copy, given that England and France were at war most of that time. As for the other set, which had been "submerged thoroughly" into the Dalrymple family library, it didn't surface until the 1970s. He prepared to cite documentation, but I yielded the point.

It was Dalrymple's 1779 pamphlet that Beaufort copied. The pamphlet hadn't achieved much currency, Cook pointed out, but Dalrymple did give a copy to Beaufort, along with a collection of his maps, in 1805, when they first met. You can see the maps and the papers, in fact, in the Library of Congress. "We know it's Beaufort's collection because it's inscribed," Cook said, cutting off further question. "And we know when Beaufort received it because at the time he received the charts and publications he used any bit of paper that he could find to construct for himself lists of the charts in the albums that accompanied the publications. And he performed these lists on the backs of unfolded letter sheets. One of those is dated November 1805, which ties it down fairly nicely."

Well, it'll do until something more exact comes along. Those same maps, by the way, were the ones on the backs of which Beaufort made some of the sketches and maps of Montevideo to which I referred during my trip there. Cook did want to keep one point central, though. The point of the numbers, he emphasized, was transmissibility: The scales Smeaton and Dalrymple designed were made so that information about the wind could be not just gathered but shared, transmitted. The point was to make the information useful. "The whole question of Dalrymple's wishing East India Company captains to use a formal wind scale was not for itself," he reiterated at the end of our cappuccino in the British Library café. "It was an adjunct to a more scientific recording of *track.* He was trying to find the best track at any season to get ships to the Indian Ocean the fastest. And so he wished to identify the prevailing weather conditions at different places.

"Mere recording is a sterile occupation," he said. "It's in order to *transmit* this data to others that the scale is made"—in Smeaton's case, to transmit directions to millers for more effective milling; in Dalrymple's, to transmit data to the company hydrographer, Dalrymple himself, who could use it to plot better courses in the future for more effective shipping.

Based on my questions, Cook had, of course, dozens of other sources he thought I should consult—Admiralty records, histories, maps, and documents—but I believe when he saw my eyes rolling backward into my head he took pity on me. He wished me well, reminded me to keep transmissibility in mind, and returned to the India Office records amid the portraits of commanders and potentates. Dalrymple himself did not exit so gracefully. A champion of Beaufort from the moment he met him, Dalrymple lasted only a few more years at the Admiralty. Slow to publish charts unless he was quite certain of their accuracy, he was eventually regarded as something of an impediment to rapid dissemination of information, hydrographic or otherwise. Stories circulated that he sometimes slept in his office, that he was short-tempered and even rude; frustration with him grew among the Lords of the Admiralty. When in 1808 the Admiralty Chart Committee demanded to see the charts Dalrymple had since 1795 retained from the captured French ship *Hougly,* Dalrymple refused to hand them over, considering them articles of science and not of war, and thus not rightfully spoils. The committee insisted, and though Dalrymple turned over the charts, he refused to share the background calculations. Forced to resign over the incident, Dalrymple apparently took it hard—he died three weeks later.

It's a shame about the wind scale—the version in *Practical Navigation* is in every way the direct forebear of the scale Beaufort wrote in 1807. And really, it seems almost certain that not only did Beaufort see Dalrymple's list of winds, but Dalrymple must at least have told Beaufort about Smeaton's windmill scale. The table is too useful, and too perfectly expressive of the ideas the Beaufort Scale eventually embraced. In *Practical Navigation,* Dalrymple even mentioned his pictures again, making sure the reader understood that the whole point of the scale was to make sure everyone knew what they were talking about: "The probable means of introducing precision, is *by shewing distinctions clearly,* which was the object I had in view." Regrettably, the book was never published. Interest in the wind scale lived on only in Francis Beaufort.

IF COOK IS THE TOP DALRYMPLE GUY, there's likewise one top Beaufort guy in the world right now, and that's Nicholas Courtney, who wrote *Gale Force Ten,* the 2002 biography of Beaufort that, in its blizzard of detail, gives a more thorough account of the scale's genesis than did the old *Beaufort of the Admiralty.* When I sought him out,

Courtney, then at work on a book about stamps, invited me to have lunch at Brooks's, his London club. I borrowed a tie.

Courtney, doing research in the Admiralty archive, had taken the suggestion of someone there to look into the life of Beaufort, whose only biography was by then more than twenty years old. He found in the man he describes as "the Good Admiral" someone fascinating, nearly forgotten, yet easy to admire. Beaufort was powerfully, almost profoundly, driven to write things down—he left tens of thousands of written records behind him. The letters and journals now in the collection of the Huntington Library; those dozens and dozens of weather diaries in the Met Office; and an almost limitless correspondence in Admiralty records leave a thorough—sometimes too thorough—portrait.

"It was a little troublesome that he was such a whinger," Courtney said, shrugging. And Beaufort was; his correspondence, especially in his early days, is filled with complaints about his treatment at the hands of the Admiralty, where his excellence was only slowly rewarded because of the "influence" system, whereby patronage got you much farther than accomplishment. Just the same, it was in the Admiralty that Beaufort found his niche and made his life. And what Courtney admired most about Beaufort was the same thing that I did—Beaufort was interested in everything, wanted to learn everything, wanted to share everything. Beaufort was, we agreed in a men's club that had been around since the eighteenth century, the last eighteenth-century man.

THE QUALITY OF HIS CHART OF MONTEVIDEO made Beaufort a valuable character in the Admiralty, and eventually he was given more surveying tasks. In fact, his survey of the southern coast of Turkey—called Karamania by the Ottomans—was one of the transforming events of his life.

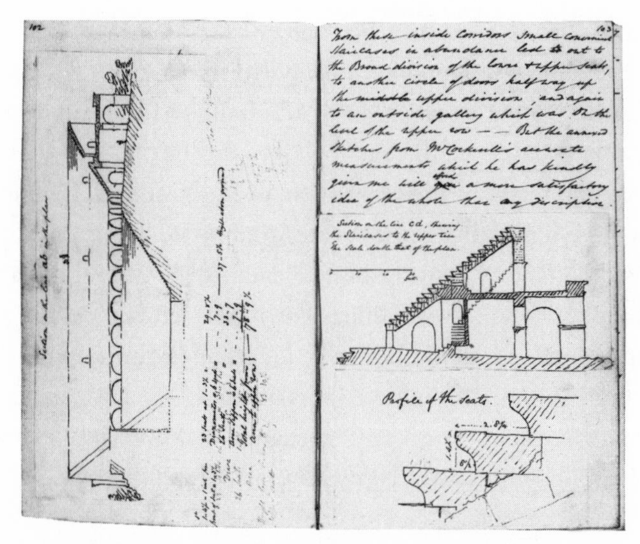

This drawing of a stadium shows the staggering attention to detail Beaufort paid in even his roughest sketches and notes.

This item is reproduced by permission of The Huntington Library, San Marino, California.

Not that it came easily, or without, as his biographer would say, some whinging. Dalrymple, already most impressed by Beaufort's surveys of Montevideo, applied to have him survey the English, Dutch, and Danish coasts (the *Woolwich* was "most admirably adapted for the Service," he wrote to the Admiralty Board, "and her Commander unquestionably as fit as any Man now living to be entrusted with the execution of the Service. I mean Captain Beaufort"), but no deal. Beaufort had already spent hours humiliating himself in the waiting rooms of the Admiralty Lords seeking advancement, and was even briefly given another command, but still, without powerful connections he couldn't gain the rank of post-captain. His letters home show him despondent: "Let me just seriously ask you," he wrote his father from the *Woolwich*, "would an immediate Post Commission at my age and with my prospects be sufficient repayment for that continually undermining of my self esteem, and thus doing every day what produces a blush on my pillow every night?"

He never lost his courage, though—during one journey back to England he discovered that a member of his crew was an escaped slave. He not only prevented Admiralty authorities from searching for the man aboard his ship; when the man did appear after the ship was under sail, Beaufort made sure he was transferred to another ship, where he would serve two years at sea and be enfranchised. On another trip, from Quebec, he arranged for his military convoy to bring back to England a widow and her children, even supplying toys for the children.

Finally, in 1810, change among the higher-ups at the Admiralty removed some of the logjam in front of him, and Beaufort was given a new command. He finally became a post-captain, chosen to command the *Fredrikssteen,* a captured Danish frigate, and in 1812 Beaufort received orders to undertake the first survey of the southern Turkish coast since ancient times.

For Beaufort, the survey was a dream come true. His entire job would be to take bearings for charts and plans, and to compare what he saw with the observations of the ancient geographer Strabo; of course he also brought with him his drawing pads and journals. He was mapping country that had been virtually ignored by Europeans since classical times, and he kept notes for a book while he traveled.

Those journals and sketchbooks, filled with breathtaking views of coastline and antiquities, are in the collection of the Huntington Museum, and they're a wonder, a marvel. Beaufort, mad for writing everything down and documenting where he was and what he saw, surpassed himself. Sketches of castles and ruined stadia and silted ports fill notebooks with insane detail. He copied down hundreds of Greek inscriptions. He explored sarcophagi. In the city of Side, Beaufort made numerous careful drawings of a crumbling amphitheater, measuring it so methodically that he could determine it would hold 15,240 people—assuming, as he wrote, that they did not cross their legs, which would take up more space.

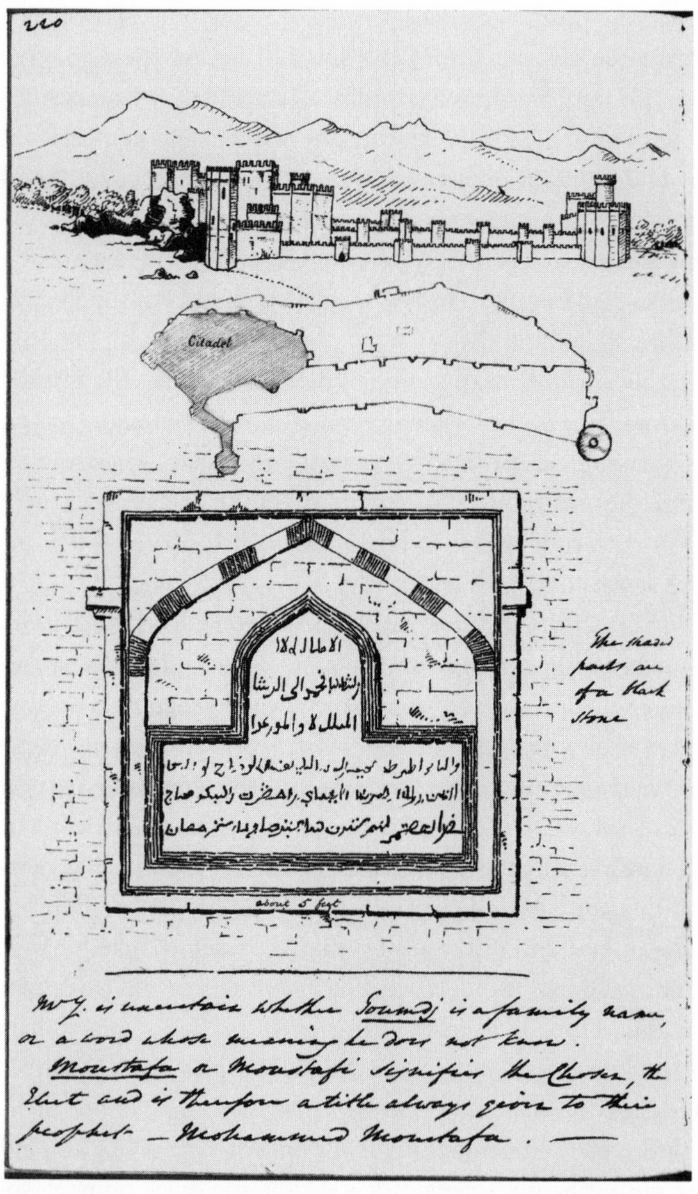

A castle, its plan, inscriptions, and notes show both Beaufort's thoroughness and the delightful juxtapositions that fill his journals.

This item is reproduced by permission of The Huntington Library, San Marino, California.

Lovely as the thorough sketches are, the quick drawings and ink washes of coastal views are equally delightful, and it is there, in paging through those journals and notebooks, that Beaufort's personality finally came clear—for me and, it turned out, for Courtney, too. Here was someone who could take astronomical measurements exacting enough to determine longitude in the middle of the ocean; who could write a plainspoken and even witty letter; who could draw like a professional. He could handle a saber as well as a quill pen, a sextant as well as a sketchbook. In the pamphlet that contains the wind scale that Beaufort copied, Dalrymple includes a section that perfectly describes life aboard ship. Discussing ways to make sketches and charts, Dalrymple describes how the artist must make do with the supplies on board: "The Medicine Chest will always afford a little Gum-Arabick to make Gum Water, if he is not otherwise provided, and a little Gambouge will give a yellow, or if he has nothing better, the infusion of a little Tobaco will afford him a brown colour. . . ." That was the environment in which Beaufort found himself. He had to be clever; to make use of everything at hand; to gather information and to communicate it. It was the environment for which Beaufort was perfectly suited, and in the trip to Karamania he found his perfect challenge.

Beaufort couldn't have met the challenge better. He not only gathered notebooks full of information and brilliant sketches on everything from antiquities to insects, but he so expertly charted the coastline that portions of his charts were still in use until 1972. On top of that, he of course kept his detailed weather journal, and he protected his crew from commonly hostile locals. He was a clever negotiator when gifts of gunpowder or cloth could gain access to an otherwise off-limits site, and he knew when to withdraw when the locals simply wanted the surveyors to leave. He managed to avoid engaging in local skirmishes, but when pressed into one—for example, when a surveying crew ran across a band of insurgents from an uprising who were about to be slaughtered—he reluctantly spirited the group to a place of safety, though he doubted their prospects were much better for his intercession.

Just the same, the survey of Karamania ended with another near-death experience for Beaufort. When, as commonly happened, a boat full of his surveyors was threatened by locals bearing muskets, Beaufort himself held them at bay with his own fowling gun, even firing over the heads of the crowd, which cowed all the mob but one. That "rascally fellow" fired from behind a rock, hitting Beaufort in the groin. He nearly died from the wound—in any case, it ended the survey.

It ended Beaufort's active career aboard ship, as well, which was fine at first, since as he convalesced he had time to write a thorough book about his exploits and to painstakingly create the charts of Karamania for the Admiralty. And yet if you read *Karamania,* his printed account of the journey, you'll find vast descriptions and beautiful sketches, but nothing that comes across as pure joy. "Opposite to these islands, and about five miles in shore, is the great mountain of Takhatlu," he says in one chapter. "The base, which is composed of the crumbly rock before mentioned, is irregularly broken into deep ravines, and covered with small trees; the middle zone appears to be limestone, with scattered evergreen bushes; and its bald summit rises in an insulated peak 7,800 feet above the sea." In another, "Situated on a gentle declivity, the lower half only of the theatre has been excavated in the ground; . . . It is shaped like a horseshoe, being a segment of a circle of about 220 degrees; or, in other words, the circumference is one-ninth greater than a semicircle. The exterior diameter is 409 feet. . . ." It sounds like what it is—surveying, not travel writing, written by someone who in spirit was more engineer than artist. Yet *Karamania* was the travel book of the season in England when it came out in 1817.

Partly this is because of the times—in 1817, just describing a strange place was entertainment enough, and Beaufort, as ever, was trying to convey the most information in the most economical way, with sketches and straightforward descriptions. But part of the reason the book caused a stir was the dawn of a new age. A year before, the British Museum had purchased the Elgin Marbles; Europe, entering the Romantic age, was going mad for antiquities, and Beaufort's book opened an entire new arena for their search—an irony in itself, since in 1799 Beaufort had been a lieutenant aboard the *Phaeton* when it had first carried Lord Elgin to Constantinople (along, it must be said, with his private secretary William Hamilton, who procured for the British a bauble called the Rosetta Stone). *Karamania,* describing antiquities hitherto unpillaged, was the spark for dozens more ventures to acquire them.

For his book, however, Beaufort, with characteristic rectitude, refused payment. He had gathered the information in the employ of the Admiralty for the public good, and so to take more money would be unseemly.

COURTNEY WAS DELIGHTED BY BEAUFORT, and it's easy to see why. Bright guy, snappy dresser, good at lots of things. He liked a challenge and liked to find a way to get things done—and get the message out. In *Karamania* he describes, for example, that distances in that part of the world are usually expressed in hours, specifically "the number of hours which a caravan of camels, or perhaps which an ordinary foot-traveller, employs in performing a journey." That in itself is a kind of Beaufort Scale of distances in Mediterranean Turkey: a unit of measure appropriate for its setting, used to convey the maximum information in the minimum words. So it's scarcely surprising that the wind scale designed by an engineer—Smeaton—and adapted by a pragmatic hydrographer—Dalrymple—had found its purest expression by this careful traveler who embodied the principles of both. Whether Beaufort was aware of Smeaton's scale remained uncertain—a member of the Royal Society, in touch with Dalrymple, he surely could have heard of or seen it; on the other hand, his letters and journals never mention Smeaton. So Beaufort's decision to organize the wind according to its graduating effects on a ship, exactly as Smeaton had done on a windmill, might be a simple case of two rigorous minds coming up with the same solution to a problem, like Newton and Leibniz independently discovering calculus. For that matter, though, in his journal Beaufort never once mentioned Dalrymple, from whom he plainly copied his first use of the scale, so his failure to mention Smeaton cannot be taken as proof he didn't know of his scale. Anyhow, I had followed the wind scale from Tycho, where I first picked up its trail, to Beaufort, where it rested. If Andrew Cook and Nicholas Courtney didn't know the exact extent of Smeaton's contribution, neither would I.

And for the moment, it didn't matter. Regrettably, with the publication of *Karamania* and its attendant charts, Beaufort, after his serious injury, retired from the navy on half-pay, though he never stopped keeping his weather diary, including its measurement of the wind.

The wind scale, after its odyssey from Smeaton, through Dalrymple, to its form as Beaufort's scale based on the sails of a ship, vanished from sight for twenty years. Still, it did remain in use, every day, by exactly one person.

Francis Beaufort.

CHAPTER 5

"Nature, Rightly Questioned, Never Lies": The Beaufort Scale, Nineteenth-Century Science, and the Last Eighteenth-Century Man

FIGURES

TO DENOTE THE FORCE OF THE WIND.

0	Calm.		
1	Light Air, - - -	Or just sufficient to give steerage way.	
2	Light Breeze, -	Or that in which a well-conditioned man-of-war, with all sail set, and clean full, would go in smooth water, from - - - - - - -	1 to 2 knots.
3	Gentle Breeze, -		3 to 4 knots.
4	Moderate Breeze,		5 to 6 knots.
5	Fresh Breeze, -	Or that to which she could just carry in chace, full and by - - - -	Royals, &c.
6	Strong Breeze, -		Single-reefed topsails and top-gallant sails.
7	Moderate Gale, -		Double-reefed topsails, jib, &c.
8	Fresh Gale, - -		Triple-reefed topsails, &c.
9	Strong Gale, - -		Close-reefed topsails and courses.
10	Whole Gale, - -	Or that with which she could scarcely bear close-reefed main-topsail and reefed foresail.	
11	Storm, - - - -	Or that which would reduce her to storm stay-sails	
12	Hurricane, - -	Or that which no canvas could withstand.	

If the above mode were adopted, the state of the wind might be regularly marked, in a narrow column, on the log-board every hour.

LETTERS

TO DENOTE THE STATE OF THE WEATHER.

b Blue Sky; whether clear or hazy atmosphere.
c Clouds; detached passing clouds.
d Drizzling Rain.
f Foggy—f̣ Thick fog.
g Gloomy dark weather.
h Hail.
l Lightning.
m Misty hazy atmosphere.
o Overcast; or the whole sky covered with thick clouds.
p Passing temporary Showers.
q Squally.
r Rain; continued rain.
s Snow.
t Thunder.
u Ugly threatening appearances.
v Visible clear atmosphere.
w Wet Dew.
. Under any letter, indicates an extraordinary degree.

By the combination of these letters, all the ordinary phenomena of the weather may be expressed with facility and brevity. *Examples*:—Bcm, Blue sky, with passing clouds, and a hazy atmosphere. Gṿ, Gloomy dark weather, but distant objects remarkably visible. Q̣pdlṭ, Very hard squalls, with passing showers of drizzle, and accompanied by lightning with very heavy thunder.

F. B.

This version of the scale might be called the classic. It was published in *The Nautical Magazine* in 1832, but this version, with Beaufort's initials typeset in the lower right corner, looks to have been handed around in support of the scale's adoption.

Met Office, National Meteorological Archive. Used by permission.

ROBERT FITZROY HAD A PROBLEM. Fitzroy was captain of the *Beagle,* a surveying ship preparing, in 1831, for a five-year Admiralty mission to survey the southern coastline of South America and, eventually, circumnavigate the globe. And Fitzroy knew that the needs of the Admiralty and of the public—and of Fitzroy himself, who wanted companionship on the journey—would be best served if he brought along a naturalist. Ship's surgeons, as the only men aboard ship who had likely been trained in science, commonly functioned as de facto science officers. But on a surveying mission like this one, Fitzroy eventually wrote in his history of the trip, everyone on the crew would be too busy for anything but the tasks at hand. He thus included among other supernumeraries an artist and an instrument maker (the Admiralty paid to feed the artist; the instrument maker, who would manage chronometers and the like, Fitzroy carried entirely at his own expense).

As for the naturalist, Fitzroy wrote, "Anxious that no opportunity of collecting useful information, during the voyage, should be lost; I proposed to the Hydrographer that some well-educated and scientific person should be sought for who would willingly share such accommodations as I had to offer, in order to profit by the opportunity of visiting distant countries yet little known." The Hydrographer to the Admiralty at that time gladly undertook the search for a likely, well-educated, and scientific person. He wrote to his friend the Cambridge astronomer Thomas Peacock, who in turn consulted the mineralogist John Henslow. Henslow said he knew someone "of promising ability, extremely fond of geology, and indeed all branches of natural history." That sounded just fine to Fitzroy, so he contacted Henslow's young man and asked whether he wished to accompany the *Beagle,* though Fitzroy could offer no pay beyond the chance for discovery and adventure. The naturalist raised questions about who would own the specimens gathered on the journey, but the Admiralty made assurances that whatever the young naturalist collected would belong to him. Fitzroy himself was uncertain of the man when they first met (a dedicated phrenologist, Fitzroy mistrusted the shape of the man's nose), but he extended an offer.

Charles Darwin accepted.

The Admiralty's hydrographer who undertook the search that put Charles Darwin aboard the *Beagle* was, of course, Francis Beaufort, who had attained the post in 1829. And if the voyage of the *Beagle* is forever remembered because of the observations Charles Darwin made thereon, for meteorologists it holds another special distinction. On the *Beagle,* Captain Fitzroy followed the suggestion of his hydrographer and expanded the ship's journal to include not only vague descriptions of the wind ("the ambiguous terms 'fresh,' 'moderate,' &c., in using which no two people agree") but specific numbers of wind force. For those force numbers he turned to a sheet Beaufort had included in the huge sheaf of memoranda delivered to Fitzroy as he prepared for the journey.

When the journals of the *Beagle* were thereafter released, they represented the first-ever official use of the Beaufort Scale.

AND IN THOSE TWO ACTIONS—providing a link in the chain that created perhaps the most influential scientific enterprise of the nineteenth century, and using the Admiralty resources to propagate the scale that bears his name—Beaufort demonstrates, at age fifty-seven, the contribution for which he is most, and most properly, remembered.

In 1831, Beaufort had been Hydrographer to the Admiralty for a mere two years, but it was the job he had been born to do. All around him, in a grand and explosive movement, modern science was being invented every day. Beaufort, the last eighteenth-century man, may not have been well suited to develop or practice the theoretical science that produced theories of evolution, electricity, atomic chemistry—the science that grew beyond the simple observational practices he had learned in his youth in the 1780s and 1790s. But he was perfectly suited—and, in the Admiralty, perfectly situated—to play the role of connector, introducer, communicator. As men around him discovered elements and created theories, Beaufort contributed by becoming a clearinghouse, a great scientific networker—the man who got the word out.

It's not a coincidence. What caught my attention about the Beaufort Scale was at first the beauty of its language, but there was something else, something powerful, about how it does its job. What the Beaufort Scale is, fundamentally, is *scientific* language. Its descriptions are beautiful, to be sure—but what they also are is distilled, thorough, complete. The Beaufort Scale, in Beaufort's form, takes the wind at sea, anywhere all over the planet—wherever a ship might encounter it—and reduces it to a format that is not only clear but quantifiable and communicable. The Beaufort Scale takes observation and turns it into information.

THAT'S WHAT WAS HAPPENING IN THE early nineteenth century. Observation was turning to data—science was becoming a profession, an identifiable pursuit with a standard pattern and method and even a recognizable practitioner.

The sixteenth through the eighteenth centuries are often called the first scientific revolution—around 1600 the use of the telescope helped Galileo understand the heavens, expanding on the work of Kepler and Tycho and Copernicus; in the early 1600s Francis Bacon firmed up the combination of observation and induction that became accepted as the scientific method; late in that century Newton propounded the theory of universal gravitation that changed the way the world understood basic physics. Much later theories—of Darwin, Einstein, and Freud, for example—are sometimes called the second scientific revolution. But between the two, especially in the late 1700s and early 1800s, the beginnings of almost all the practices of modern science occurred.

Benjamin Franklin, using the newly created electricity-storing

Leyden jar, understood the fundamentals of electricity around 1750, publishing his famous description of the kite in 1752; by the 1780s Galvani was making frogs' legs twitch with electricity, and by 1800 Volta was making batteries. By the 1820s Faraday was conducting experiments in electromagnetism, and the electrical age had begun in earnest.

As the Industrial Revolution gathered momentum, metals became tremendously important, so mineralogists paid more attention to the Earth, and in 1795 James Hutton's *Theory of the Earth* introduced the concepts of modern geology—that silt and sand deposited through millennia beneath ancient seas fused into rock by pressure and heat, that the sedimentary layers provide a historical geological record (though the theory of continental drift that we take for granted now was not proposed until the early twentieth century).

In 1759 a comet previously seen in 1607 and 1682 returned, just as Edmund Halley had predicted in his calculations based on Newton's gravitational theory. The comet's return removed any doubt about whether Newton's conclusions about the motions of the planets, and hence of objects in space, were correct. Henry Cavendish discovered hydrogen in 1766; Daniel Rutherford separated nitrogen in 1772; and in 1774 Joseph Priestley discovered what he termed "dephlogisticated air," which Antoine Lavoisier called oxygen. Before long those beginnings of chemical understanding led to the closer observation of plants and the understanding of photosynthesis. Lavoisier, following the work of Priestley, also recognized that chemical substances interacted in predictable and expressible ways—hydrogen and oxygen, for example, predictably made water, and did so in consistent proportions, giving rise to the idea that chemical processes could be expressed as balanced mathematical equations, creating the modern practice of chemistry.

The improved understanding of chemistry led to an improved

understanding of the building blocks of matter. The concept held by Aristotle and the ancients that everything was made up of combinations of four basic elements—earth, air, fire, and water—still had currency in the seventeenth century; in 1661 Robert Boyle wrote *The Sceptical Chemist,* criticizing those who followed Aristotle (or the medieval alchemists who proposed such basic elements as mercury and salt). Boyle suggested that elements were many, though he didn't know what they were, or exactly how to find them. A hundred years later, Priestley and Lavoisier understood that certain substances (like hydrogen and oxygen) had specific properties, including weights, and thus careful measurement could determine the composition of complex compounds. In 1808 John Dalton recognized that those properties, and the predictable proportions of the elements in differing compounds, suggested that matter was composed of elemental particles (he called them atoms, just as Democritus had done around 400 B.C.) and that the atoms of each element shared such characteristics as weight and behavior. Dalton thus created the first organized chart of the elements based on atomic weights, though the periodic chart, as we know it now, was still sixty years in the future. (In 1813 the Swedish chemist Jons Berzelius began referring to them by their first initials: *O* for oxygen, *H* for hydrogen. It stuck.)

That missing periodic chart, though, brings up an important point. A good deal of what was happening in the sciences wasn't simply leaps forward in understanding; it was improvements in organization and expression. In 1709 Gabriel Fahrenheit had made an alcohol thermometer; in 1714 he developed a mercury thermometer that became a standard, with its zero at the lowest point reachable in laboratories at the time and 96 chosen, somewhat arbitrarily (possibly because it was easily divisible by several numbers), as body temperature. In 1742 Celsius did the same thing, only slightly more conveniently, with zero at the freezing point of water and 100 at its boiling point. That is, the

Celsius scale was better organized and so more useful, and it's no surprise that scientists almost exclusively use it. As scientific enterprise expanded, observations needed to be not only made but communicated, so common expressions needed to be organized, agreed upon, and shared. Linnaeus, after all, had published in 1758 the tenth edition of his *System Naturae,* which created the Latinate taxonomic system still used in the biological sciences today. Less momentously, if you turn on the Weather Channel today, you'll hear about cumulonimbus or nimbostratus clouds—names assigned in 1804 by the amateur meteorologist Luke Howard.

Travel was becoming easier; improved shipping spread communication throughout the world. And as communication became more regular and even constant, standard ways to share information were required. In the Revolutionary spirit of reinventing everything from the ground up, the French in 1792 designed the metric system, which finally offered a series of standard measures tremendously useful to the growing sciences; Lavoisier, the chemist who gave oxygen its name, was part of the committee that figured it all out.

Not every advancement in metric standardization went well, of course. Those Revolutionary French occasionally took things too far. Don't forget, in 1793 they decided it made sense to rename the months (Fructidor, Thermidor, and so forth), and to make them all equal in length at thirty days, and divided into three equal ten-day segments. They forced that on the population, and it lasted a dozen years or so, until the pragmatic Napoleon returned France to traditional months and weeks in 1806. Things in fact got sufficiently out of hand during that period that the French actually executed, among others, Lavoisier: "The Republic has no need of scientists," the tribunal supposedly said.

The French took their calendric reorientation too far, but the calendar did need fixing. It wasn't until 1752 that Britain finally adopted the Gregorian calendar the rest of Europe had adopted in 1582: Britons went to bed on September 2 and woke up the next morning on September 14. There were riots, as people thought they'd somehow been cheated out of eleven days, but in any case all of the West was finally on the same calendar page.

This standardization of measurements, organization of perceptions, is not merely natural—it's actually fundamental. Psychologists of perception will tell you that categorization is the fundamental perceptive activity: It's what enables organisms to function. One of these things is not like the other—why, and what does that tell us? All organisms can make the most basic distinctions—between food and not-food, danger and safety, light and dark, same-species and not-same. But only people can use language to make the highly complex categorizations of, say, animals or physical forces, or however many different kinds of quarks there are now, putting them in separate piles and naming the piles. It's how we proceed; it's how we communicate. Organization into categories is, at bottom, human.

The very first thing Adam is shown doing in the Bible, after all—the very first human action ever described—is naming the beasts, the same thing Linnaeus did after him, the same thing Howard did with the clouds and Dalton did with the elements. The same thing Beaufort did with the winds. Not for nothing is so much of our understanding, so much of our science, based on scales, on classification systems. It's how we reduce the complex into the understandable, the chaotic into the ordered, the incomprehensible and nameless into the structured and named. Mankind is an organizing species. Our science makes us so.

THE VERY TERM *SCIENTIST,* by the way, is itself a pilgrim of Beaufort's generation. "We need very much a name to describe a cultivator of science in general," wrote one William Whewell in 1840. "I should incline to call him a scientist." Whewell was a British astronomer known for, among other things, bringing the developing European standard mathematical symbols to British mathematicians and for coining many words we now take for granted in science—*physicist,* for example, and *Miocene* and *Eocene* in historical geology and *ion* and *cathode* in electricity.

Until Whewell, scientists were called natural philosophers, which demonstrates perfectly who they largely were: gentlemen, philosophers. Those few educated men who could afford to spend time and money pursuing intellectual questions did so; everyone else was busy scratching their livings from the earth. Science—natural philosophy—was something rich, leisured people did. A glance at the *Philosophical Transactions* of—to pick a decade at random—the 1740s gives a good example. "A Letter from Edward Nourse, F.R.S. Surgeon to St. Bartholomew's Hospital. . . . Giving an Account of Several Stones Found in Bags Formed by a Protrusion of the Coats of the Bladder, as Appeared upon Opening the Body of One Mr. Gardiner"; "Extract of a Letter from Mr. Christopher Mason, concerning a Fire-Ball Seen in the Air, and a Great Explosion Heard, Dec. 11. 1741"; "An Account of an Extraordinary Case of the Bones of a Woman Growing Soft and Flexible; Communicated to the Royal Society by Mr. Sylvanus Bevan, F.R.S."

These are fascinating things seen or retold: Hey, you guys! I saw the coolest thing! Some attempt at explanation commonly accompanies the description, but it's basically pure description. By 1832, when Beaufort was running the Hydrography Department and the *Beagle* was carrying Charles Darwin to his destiny, articles in the *Philosophical Transactions* had an entirely different cast: "On the Tides," by John William Lubbock—part of a worldwide effort in gathering data that enabled people to learn, for the first time, that tides were connected worldwide; "On an Inequality of Long Period in the Motions of the Earth and Venus"; "Experimental Researches in Voltaic Electricity and Electro-Magnetism." This is another matter entirely—this is experimental science, organized, reproducible, communicable. It's science we'd recognize today. It was a period of organization, of clarification and classification. Scientists had begun to understand the movement of the wind a century before, yet the final understanding that air moved around the planet in streams and masses was still a hundred years off. (Observers as early as Benjamin Franklin had recognized that storms moved, and not necessarily in the direction of winds felt on the ground. But only in the 1910s did meteorologists finally recognize that mid-latitude storms—the kind most of us are most likely to experience—move not so much in discrete streams of air as with the moving boundaries between large masses of air of differing temperature and pressure. Thus a mass of cold air advances on a mass of warm air, and the turbulence at their border causes storms, which move as the air masses do. The Norwegian meteorologist Vilhelm Bjerkness developed this theory during World War I, so he naturally called the unsettled borders of these huge air masses "fronts.")

What was understood in Beaufort's time, however, was that wind blew, and it blew everywhere, and the best way to understand it was to gather data. That heaping up of information, of repeated observation, was key—from that, the order could eventually be induced. It was the Baconian method in its purest—and perhaps final—manifestation. As Francis Beaufort wrote in 1809 to his friend Richard Lovell Edgeworth, there were a thousand Admiralty vessels afloat, each depositing several logbooks per year into the Hydrographer's office, each containing hourly observations of the wind and weather, "spread over a great extent of ocean. What better data could a patient meteorological philosopher desire?"

IF DESPERATE TIMES CALL for desperate measures, remarkable times reward remarkable men. Francis Beaufort was the perfect man to function as clearinghouse in the middle of this scientific organizational flowering.

His biographer told me that Beaufort was like an ad for the auto club, in which someone calls for help. "They ask if they can fix it—and the person says, 'No, but I know someone who can.' That was Beaufort. If he didn't know the answer to your question, he knew the right man to get it for you." He was prepared, connected, and ready for action. When Fitzroy was looking for a naturalist, Beaufort didn't happen to know Erasmus Darwin's grandson, but he sure knew how to find him. Beaufort was connected with every scientific society and community—not only was he offered secretaryship of the Royal Society (he declined), he was a member of the two other most important scientific societies in Britain, the Royal Astronomical Society and the Royal Geographical Society. (He also eventually was a member of the Royal Society for Literature, l'Académie des Sciences de l'Institut de France, the Royal Irish Academie, and the United States Lyceum.) Among the names in a list in his papers titled "Remarkable people constantly at the House" are people like Charles Babbage, one of the earliest conceivers of the computer; Peter Mark Roget (he wrote the thesaurus); presidents of the Royal Society; admirals; and the surveyor-general of India, George Everest.

Beaufort's survey of Montevideo had got him the attention of the Admiralty; his survey of and book about Karamania had made him famous. He was elected to the Royal Society, and he began meeting and corresponding with the great scientists of the age. He spent much of his time involved in mostly unsuccessful business ventures, investing in, among other things, a slate quarry (the building industry immediately slumped) and the auction of a library of one of his long-dead Beaufort relatives (the auction didn't even cover the costs of transportation). Notably, though, in 1826 he became involved in the Society for the Diffusion of Useful Knowledge, a group of men interested in getting maps and books into the houses of the growing middle class.

When Thomas Hurd, the hydrographer who had succeeded Dalrymple, died in 1823, Beaufort applied for his job, but the Admiralty was still a political place, and Beaufort was passed by in favor of William Parry (known as "Parry of the Arctic" to the British). In 1817 the Admiralty Hydrographic Office under Hurd had founded the Corps of Surveying Officers and built six new boats specifically as surveying ships (one was the *Beagle*). But whereas Hurd had worked hard at expanding the charts available to British captains, Parry showed a great deal more interest in his own journeys to the Arctic than in those of his hydrographers, and the department languished.

So by 1829, when the Admiralty had to replace Parry, Beaufort was the obvious choice. With his excellent skills and scientific connections Beaufort was offered the job, instantly took it, and thereafter turned the Admiralty Hydrographic Office into the greatest of its kind in the world. Histories of hydrography have called Beaufort's tenure—it lasted until 1855, when he finally retired at age eighty-one—"the High Noon of hydrography." Though Beaufort's specific qualifications—his penchant for detail, his obsessive documentation, his love of organization and communication—account for a great deal of this success, some of it is due simply to an unprecedented intersection of advancing scientific knowledge; improvements in transportation, tools, and technology; and the overwhelming sea power of the British Empire. The Admiralty Hydrographic Office was simply the natural repository for the information these ships brought back.

Collecting information from ships was hardly a new idea. When Dalrymple was Hydrographer to the Admiralty, he had created a form for the books of "Remarks on Coasts, Harbours, & C." that captains were directed to fill out regarding any port that they "may visit from time to time." Included are, of course, thorough physical descriptions (latitude, longitude, altitude of high points); places to anchor, find wood and water and provisions; and information about fortifications and trade. Finally, though, the form includes a special area for "Religion, Government and Disposition and Language" of the inhabitants. That is, in addition to information that would prepare the Admiralty for military encounter or the merchant marine for trade, the Admiralty wanted to know—pretty much anything else the ships could find out.

The Admiralty had been in the finding-things-out trade for a long time, of course—what were Cook's voyages but a good long reconnaissance?—but only at around the time of Beaufort's tenure did the gathering of scientific information emerge as a central project on any voyage. In fact, not until 1831 did the Admiralty create a Scientific Branch, including the Hydrographic Department and, among other offices, the Chronometer Office. (The Chronometer Office, among other things, organized the first "time ball" in the world. A certain Captain Wauchope of the Royal Navy suggested that perhaps the best way to help captains rate their chronometers—that is, to set them and make sure they were running accurately—would be to make a consistent visual signal at a specified time each day, much as morning or noon cannons were sometimes used. The Admiralty tried out his plan in 1829 at Portsmouth, and in 1833 the Royal Observatory at Greenwich began dropping a red ball from a windvane atop its main building each day at 1:00 P.M. Captains in port could train a telescope on the ball and thereby set their chronometers. The system worked so well that time balls spread throughout the world, and the Greenwich ball still drops daily. The first ball dropped on New Year's Eve in New York, by the way, in 1907.)

Above all, this Admiralty spirit of scientific enterprise required observation—the steady accumulation of data that Beaufort and other "patient . . . philosophers" could then study. Much of this science supported navigation, naturally, but there was more. Beaufort certainly wanted his wind scale used to gather data that would improve nautical effectiveness, but he also wanted it to supply data for purely scientific enterprises—enterprises whose goals he couldn't yet imagine.

This was a leap forward for the wind scale—and for the Admiralty. Smeaton designed his scale to help people use windmills, nothing more. Dalrymple adapted it to improve sailing. But Beaufort's scale, finally, was designed to gather, store, and communicate data, for whatever purposes it might eventually be used. This was a whole different undertaking, and it's a true contribution to science.

This is described beautifully in Beaufort's instructions to Fitzroy; it's exemplified in some of the information Fitzroy and Darwin brought back. But it's distilled to its essence in a book that in 1849 Beaufort produced as hydrographer, a book called *A Manual for Scientific Enquiry* that, even more than the Beaufort Scale, might be the perfect expression of Beaufort's questing and observant spirit.

IN FITZROY'S PUBLISHED ACCOUNT OF HIS JOURNEY, Beaufort's 1831 instructions went on for seventeen pages. As comparison, in 1826, when the *Beagle* and its companion, the *Adventure,* left for a previous journey, the Admiralty Board wrote instructions to the commander of the voyage, telling him when to leave, where to stop on the way, and thence to survey the coasts "in such manner and order, as the state of the season, the information you may have received, or other circumstances, may induce you to adopt. . . . You are to continue on this service until it shall be completed." They were done in less than five pages.

And that was about it.

With Beaufort at the department's helm, voyages of discovery ("an expedition devoted to the noblest purpose, the acquisition of knowledge," Beaufort said in his instructions) received a lot more attention. Beaufort presumed that the original instructions to the expedition to "avail yourself of every opportunity of collecting and preserving Specimens of such objects of Natural History as may be new, rare, or interesting" continued in force, and he then went on in tremendous detail about how to break the initial run to Rio de Janeiro into four parts, and what to survey on the way; what to look for in the seemingly uninteresting coast south of Montevideo ("the more hopeless and forbidding any long line of coast may be, the more precious becomes the discovery of a port which affords safe anchorage and wholesome refreshments"); which rivers to survey; and generally in what order to do so, though he pointed out that it would be ridiculous of him to try to plan the journey from his office, instead preferring "to rely on the Commander's known zeal and prudence."

He notes that the charts should include "views of the land," though "plain, distinct roughs . . . on a sufficiently large scale to show the minutiae of whatever knowledge has been acquired, will be documents of far greater value in this office . . . than highly finished plans, where accuracy is often sacrificed to beauty." He especially urged artists to forgo detail in hills, which "are so often put in from fancy or from memory after the lapse of months, if not years." The inshore sides of hills, he points out, can't even be seen from the ship and so "must always be mere guess-work, and should not be shown at all." That Beaufort felt the need to remind his surveyors not to try to represent what they could not see tells a great deal about what he expected to learn about what they *could* see. Beaufort expected the details.

He wanted information about the monsoons; about the mountains; about the tides; about the deflection of the compass; about ocean currents. "If a comet should be discovered while the *Beagle* is in port," there's a procedure to be followed. Sea and land temperatures should be noted, and of course he wanted the wind observed. Nothing should be wasted, nothing ignored—everything documented, stored, distilled for some time when it might be useful.

And that's just what the Admiralty ships were bringing back, too. A glance at the kinds of things the *Beagle* found is instructive. The volume about its first voyage, when it sailed with the *Adventure*, includes among navigational information and travel details that the tide is especially tricky near Port Santa Elena; that the natives were cheerful after receiving medals, especially minted for that purpose; that the crew started a cooking fire onshore that "transmitted to the stubbly grass" and eventually burned fifteen miles of shoreline to a depth of seven miles inland, the smoke so thick that "it quite impeded the sun." Among other details are the "dimensions of a native," presented in tabular form: "From the top of the fore part of the head to the eyes, 4 inches . . . to the tip of the nose, 6 inches" and so forth.

It's an irresistible image: "Greetings old boy—Fitzroy here, of the *Beagle.* Dreadfully sorry we burned up your country, but here's a nice medal for you. Don't mind if we just have a little measure of your head, do you? There's a good chap, cheerio then, carry on." It's just marvelous; this was exploring in the purest sense imaginable. Who were the people? What did they say and do? What did they eat, drink, hunt? Where could you find water, trees for masts and fuel, tribes who would trade food for knives and trinkets? The natives would signal to the ship by building a fire—the ship would signal back by raising the flag. Cultures were meeting, sniffing each other out, and bringing information back—the most basic information: "They subsist chiefly on fish, which they catch either by diving, or striking them with their darts. They are very nimble afoot, and catch guanacos and ostriches. . . . They are innocent, harmless people."

When Fitzroy became captain and the *Beagle* sailed alone, the voyage produced three volumes. The first was a thorough recounting of the journey from Fitzroy's journal. Second was an appendix volume, containing among other things the weather journal that for the first time listed wind speeds according to the Beaufort Scale. The appendix also included copies of the ship's correspondence and a host of other interesting things—copies, in Latin, of passages of the history of the journeys of Americus Vespucius to clarify Fitzroy's claim that Vespucci could never have sailed farther south than the Rio de la Plata; coastal observations (what could be found in each port); remarks about the wind and tides in different spots; the basics of several native languages; and, of course, the phrenological observations that interested Fitzroy: "Taking a general view of the head, the Propensities (the organs most exercised by a barbarian) are large and full; . . . the Intellectual organs, which are chiefly used by man in a civilized state, are small."

Darwin's observations form the final volume of the narrative, and they constitute, of course, one of the treasures of scientific literature in English. Starting off with thanks to Beaufort and the Admiralty and proceeding with a straightforward diary of what he did and saw from July 1832 to October 1836, the journal is a treasury of straightforward descriptive prose. "The neighbourhood of Porto Praya, viewed from the sea, wears a desolate aspect," Darwin begins. "The volcanic fire of past ages, and the scorching heat of a tropical sun, have in most places rendered the soil sterile and unfit for vegetation." The narrative never strays far from that tone of both vivid description and borderline toneless recounting. You can easily find statements that raise eyebrows in the light of history, of course. In October 1835, Darwin describes rodents of the Galápagos: "There is also a rat which Mr. Waterhouse believes is probably distinct from the English kind," he says, "but I cannot help suspecting that it is only the same altered by the peculiar conditions of its new country." From there it's hard not to cast ahead to the more general conclusions Darwin eventually drew.

In many ways all the volumes from the *Beagle* bear a striking resemblance to Beaufort's own *Karamania,* which had the same simple scientific purpose: to recount a journey, to describe hitherto unseen places, people, flora, and fauna for an audience that was already fascinated and needed no artificial narrative to keep it interested. Darwin shares tales of trips into unexplored territory: "It would be difficult to imagine a scene where [mankind] seemed to have less claims, or less authority. The inanimate works of nature—rock, ice, snow, wind, and water—all warring with each other, yet combined against man—here reigned in absolute sovereignty"; unusual animals seen: "A spider which was about three-tenths of an inch in length . . . while standing on the summit of a post, darted forth four or five threads from its spinners. These glittering in the sunshine, might be compared to rays of light. . . . The spider then suddenly let go its hold, and was quickly borne out of sight"; and new people encountered: "Soon afterwards we perceived by the cloud of dust, that a party of horsemen were coming towards us; when far distant my companions knew them to be Indians, by their long hair streaming behind their backs. The Indians generally have a fillet round their heads, but never any covering; and their black hair blowing across their swarthy faces, heightens, to an uncommon degree, the wildness of their appearance."

These, Darwin seems to know, are narrative enough on their own. So Darwin often writes like Fitzroy, like Beaufort: just the facts, but the facts in words that convey real information. The aeronaut spiders in Darwin's description didn't dart forth "a few" threads, they darted "four or five"—Darwin obviously watched those spiders for a while and described exactly what he saw. In fact, the spiders could almost function as a stop on the Beaufort Scale itself: One presumes that at force 1 or 2, the spiders wouldn't waste their time trying such a meager breeze, and that by force 6, they probably just take cover. Perhaps a nice force 4, then, could include Darwin's description: aeronaut spiders "quickly borne out of sight." It's not impossible that Darwin, reading the voluminous Admiralty instructions to Fitzroy, encountered the wind scale and took a few descriptive cues from the hydrographer.

BEAUFORT'S DESCRIPTIVE STYLE TRULY PEGS HIM as the last eighteenth-century man. Though the *Philosophical Transactions* showed that by the 1830s science was progressing into the theoretical and mathematical methods with which we are more familiar now, the observational science undertaken by the scientists aboard Admiralty ships bore a significant hangover of the simple natural philosophy of the late 1700s, when Beaufort himself, in his teens and twenties, was being educated. In some ways the scientific information the Admiralty ships were bringing back represented the last flowering of eighteenth-century science—the final expression of purely observational natural history before just about everything had been seen, "reflecting more the attitudes of Sir Joseph Banks' era," according to Friendly's biography. (Sir Joseph Banks was the naturalist who accompanied Captain Cook and eventually the longtime president of the Royal Society during its heyday of anecdotal descriptive science—in *Mutiny on the Bounty,* when Charles Laughton as Captain Bligh promises "I shall commend your industry to Sir Joseph Banks," that's who he means.)

And if anything in Beaufort's work at the Admiralty expressed that spirit, it was, even more than the Beaufort Scale, a remarkable book called the *Manual of Scientific Enquiry,* which the Hydrography Department published in 1849. Only someone as connected in the scientific community as Beaufort could have brought this project off; and only someone with Beaufort's eighteenth-century viewpoint would have conceived it.

Its full title describes the book: *A Manual of Scientific Enquiry; Prepared for the Use of Officers in Her Majesty's Navy; and Travellers in General.* The preface, in the form of a memorandum from the Lords

Published by
Authority of the Lords Commissioners of the Admiralty.

A MANUAL

OF

SCIENTIFIC ENQUIRY;

PREPARED FOR THE USE OF

OFFICERS IN HER MAJESTY'S NAVY;

AND

TRAVELLERS IN GENERAL.

ORIGINALLY EDITED BY

SIR JOHN F. W. HERSCHEL, BART.

Third Edition,

SUPERINTENDED BY THE

REV. ROBERT MAIN, M.A., PRES. R.A.S.

LONDON:
JOHN MURRAY, ALBEMARLE STREET,
PUBLISHER TO THE ADMIRALTY.
1859.

The right of Translation is reserved.

The Manual of Scientific Enquiry,
a distillation of nineteenth-century methods of observation.
Graduate Library, University of Michigan.

Commissioners of the Admiralty "Relative to the Compilation of a Manual of Scientific Enquiry, for the use of Her Majesty's Navy," reads like the table of contents of Beaufort's brain (Friendly, in fact, says the memorandum is written in "prose suspiciously like Beaufort's"; but though it's certainly plausible that Beaufort drafted the memo for his own project, there's no evidence that he did so).

"It is the opinion of the Lords Commissioners of the Admiralty," the memo begins, "that it would be to the honour and advantage of the Navy, and conduce to the general interest of Science, if new facilities and encouragement were given to the collection of information upon scientific subjects by the officers, and more particularly by the medical officers, of Her Majesty's Navy, when upon foreign service; and their Lordships are desirous that for this purpose a Manual be compiled, giving general instructions for observation and for record in various branches of science."

So far, so good, but then Beaufort the eighteenth-century man truly emerges: "Their Lordships do not consider it necessary that this Manual should be one of very deep and abstruse research. Its directions should not require the use of nice apparatus and instruments: they should be generally plain, so that men of merely good intelligence and fair acquirement may be able to act upon them; yet, in pointing out objects, and methods of observation and record, they might still serve as a guide to officers of high attainment."

For twenty-first-century readers, accustomed to thinking of science as something undertaken by people in white coats in laboratories, a statement like this is almost breathtaking. This is a manual of science, to be sure, but that new word "scientist" doesn't limit it to a special class of people. It's a manual designed to enable a reasonable man of average curiosity and smarts—"of good intelligence and fair acquirement"—to use the materials he finds at hand and bring back such information as he can for queen and country. It's delightful, and of course it goes on. Beyond exact science, their lordships would appreciate hearing "Reports upon National Character and Customs, Religious Ceremonies, Agriculture and Mechanical Arts, Language, Navigation, Medicine, Tokens of value, and other subjects," though on those diverse topics they have no specific suggestions, choosing merely to encourage rather than instruct their observers.

The memo lists the book's topics, and the authors of each section demonstrate the breadth and strength of Beaufort's scientific connections: The general man of science Sir John F. W. Herschel wrote about Astronomy and edited the book; presidents of the Royal Society wrote the sections on Magnetism and on Geography. The man responsible for the geological survey of England wrote on Mineralogy. There were chapters on Meteorology and Botany, on Statistics, Tides, and Zoology, and of course one on Hydrography.

Charles Darwin wrote the section on Geology.

THE *MANUAL* WAS A WHO'S-WHO of science in the mid-nineteenth century—but again, what's most remarkable about the book is neither its contributors nor its organization: it's the way it instructs its readers. In the chapter on hydrography, for example, Rear-Admiral F. W. Beechey gives careful instructions on how to describe features of an approaching coast: describe anything white, which will stand out against the dark background of the coast; do not point out white objects with only sky as background, however—they won't be sufficiently visible to navigators. Above all, "Always bear in mind that no description can equal a tolerably faithful sketch, accompanied by bearings. . . . Always write under the sketch the name of the place, and especially the native name, if you can possibly learn it." Highly practical stuff, and stuff designed not just for the trained hydrographer.

In the section on earthquake phenomena, Robert Mallet starts with the etymology of the word *seismology* and then explains how the shaking of the ground in earthquakes moves outward from the epicenter in waves. Of course one wishes to measure the speed of those waves, and Mallet makes a brilliant suggestion. Since the earthquake will knock things over, all an observer needs to do is attach a light cord to the pendulum of his grandfather clock, drilling (and greasing) holes through the sides of the clock case to get the cord out. Then the ends of the string can be attached to a narrow log stood on end. When an earthquake comes, it will knock over the log, which no matter which way it falls will pull the cord, restrict the pendulum, and thereby stop the clock at that second. Assuming the clock was set to Greenwich time, people surrounding the earthquake need only compare at what time their clocks stopped to have a record of the speed and direction of the seismic wave.

No special equipment needed; no special knowledge needed. Just a clock and a little bit of ambition, and anybody could be a seismologist—even a seismographer. The *Manual of Scientific Enquiry* is a book about science after there were "scientists" but before science became a spectator sport.

It's filled with that spirit—filled with advice that helps scientists but is also useful for, as its title explicitly states, "Travellers in General," which, really, means everyone. Section after section gives general suggestions that not only serve those observing the particular scientific process covered in that section but actually should always be followed by any enterprising traveler anxious to increase understanding. For example, in the section on Geography, W. J. Hamilton tells his readers that "it may be, perhaps, expedient to mention a few general points which should be constantly borne in mind as the basis of all observations, inasmuch as without them, all individual remarks, however carefully made, must be desultory and unsatisfactory.

Art. XIII. EARTHQUAKE PHENOMENA. 345

case, at *a b*, at the level of the lowest point of the pendulum-bob, and in the plane of its vibration; round off the edges of these holes, and grease them.

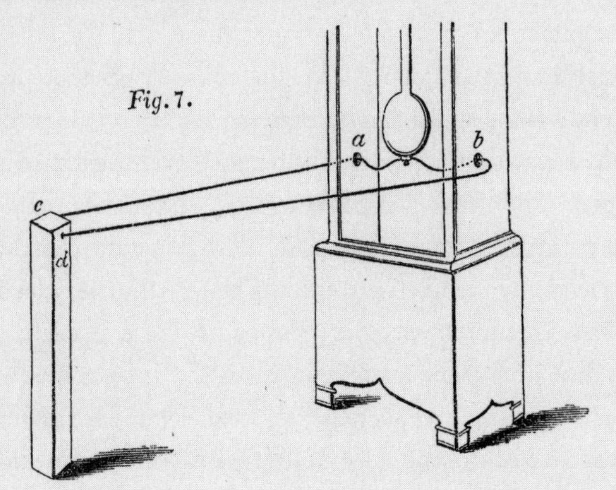

In the centre of a piece of fishing-line or stretched whip-cord make a loop and pass it round the screw or other lower projection of the pendulum-bob; pass the two free

Seismology in *The Manual of Scientific Enquiry:*
An earthquake knocks over the post, stopping your clock.
Compare yours with others all over the
country, and you can measure speed and direction of seismic waves.
Don't forget to make sure the clock is set accurately!
Graduate Library, University of Michigan.

"Most prominent amongst these general points is the necessity of acquiring a habit of writing down in a notebook, either immediately or at the earliest opportunity, the observations made and information obtained. . . . A thousand circumstances occur daily to a traveler in distant regions, which from repeated observation may appear insignificant to himself, but which may be of the greatest importance to others, when brought home in the pages of his note-book. . . . It is dangerous to trust much to the memory on such subjects." Hamilton—once president of the Royal Society—also urges observers to avoid overgeneralizing: Just because it's rocky here doesn't mean it's rocky on the other side of the hill. He reminds his observer to carry a compass, too, saying in summation that an observer should "acquire the habit of never quitting his ship without his note-book and pencil and his pocket-compass, and although at times it may seem irksome to have to remember and to fetch these materials, the traveler, if he acquires the habit of constantly using them with readiness, will never have reason to regret the delay of the inconvenience . . . in providing himself . . . with such useful companions." Another life lesson. A handbook of preparedness in a few sentences.

And in the section on earthquakes, Mallett makes another point: The observer should not content himself with hearsay—he must see for himself. When questioned, people might at least exaggerate or even, for any number of reasons, prevaricate. That's why the observer should use his own senses and take his own observations. For, as Mallett says, "Nature, rightly questioned, never lies."

"NATURE, RIGHTLY QUESTIONED, NEVER LIES." That quotation entirely sums up a book, an era, a world, a way of living.

One hundred fifty years before, people thought there were questions that simply could not be answered. Defoe, in *The Storm,* had said people could simply never know why the wind blew—in fact, too much questioning would finally throw Mother Nature herself into a rage: The answer "is not in Me, you must go Home and ask my Father." Now, whether it's nuclear weaponry, cloning, or the Human

Genome Project, we often wonder whether we are learning things we should not know.

No such uncertainty then—at that moment, Beaufort and the phalanx of freshly named "scientists" knew: "Nature, rightly questioned, never lies." They feared knowing neither too little nor too much; it was all a matter of finding the questions, and the rest would come almost as obligation. They believed that if you were patient, thorough, and careful, you could eventually figure out everything.

Oh yeah—patient, thorough, careful, and above all *observant.* The *Manual,* written not only for scientists but for men of "good intelligence and fair acquirement," for "Travellers in General," urged everyone to for goodness' sake carry around a pencil, a notebook, and a compass, so you can know where you are and remember what you've seen. Beaufort, of course, did so. The Huntington Library has his little pocket diaries, which he kept for jotting down not just appointments but observations—with whom he dined, whether they said anything interesting ("shocked to find her a total unbeliever, almost an atheist," he said of one dinner companion).

It's the best advice any book has ever given—carry a pencil and paper because you might notice something—and finally, this delightful summation enabled me to understand Francis Beaufort and why I liked him so much. The more I learned about the times surrounding him, the more I saw that he fit his times like a key into a lock. It came down to the simple thing his biographer Nicholas Courtney and I had discussed:

Beaufort was the world's greatest polymath. He was a cliché, the very model of a modern major general, knowledgeable in matters vegetable, animal, and mineral. It was all worth Beaufort's attention—from the bottom of the sea to the stars, from the Earth's magnetic variation to the tides to the Northwest Passage to the new animals

Darwin or someone else might stumble onto (say, for example, the gorilla—heard about but not discovered, if you can believe it, until 1847). Drifting through Beaufort's notebooks is like sitting next to the best dinner companion imaginable, a man fascinated not by himself but by the world.

Examples fill his papers in the Huntington. One artifact there is Beaufort's commonplace book. Such a book would have been standard in Beaufort's time—it was a collection of interesting things its owner encountered, overheard, or read; with no photocopiers, people just wrote things down. Beaufort's, like most, has an index at the back; though the book is labeled "1806–1825," it includes material from his earlier days (some pieces are labeled from the 1790s). The book is filled with items like "the uses of cobalt for staining glass"; an item clipped and pasted in from the *Cork Advertiser* of 8 October 1807, which Beaufort has headed "Method of taking bees" ("In the dusk of the evening, when the bees are quietly lodged," the article says, "approach the hive, and turn it gently over."). From the *Dublin Journal* of December 1797, Beaufort has copied in a list of the populations of the great cities in the world (London was estimated at 800,000).

There's a spurious etymology of *assassination,* an assessment of the effect of the sun on barometers, a joke about the sacredness of the color green to Ottomans, a description of how to carry grapes, the best method to capture a wild male elephant (use female elephants as bait; it's listed in his index as "elephants, taking").

Whether it was taking bees or taking elephants, Beaufort was interested in *everything.* And the *Manual* represented that interest. I found myself thinking that Beaufort was a person with whom you'd just never run out of things to talk about—and that if you ended up on a ship for a voyage of several years, you wouldn't regret having with you the *Manual of Scientific Enquiry.* It's like the Beaufort Scale writ large:

Pay attention to what you see, jot it down, and find the clearest way to communicate it. What you see is important. So is writing it down.

THE *MANUAL OF SCIENTIFIC ENQUIRY* went into several editions; by the third, in 1859, there were fifteen chapters, including medical statistics and ethnology, and a section on tides written by William Whewell, the man who had coined the term *scientist.* The book was listed as an Admiralty publication until 1913. And if you look in the Meteorology section of the *Manual,* you find, of course, the Beaufort Scale. It's still not known as such (it's called here "Admiral Beaufort's system of abbreviations"), but it implies that by 1849 the scale had become common and even connected by name with Beaufort.

In fact, that's exactly so. In 1832 the Admiralty had begun publishing the *Nautical Magazine,* used at first to provide regular updates to Admiralty Charts. Those updates, tremendously important, soon spun off into their own publication, the *Notices to Mariners,* which left the *Magazine* free to provide other items of more general interest to seamen. In its first year of publication, Beaufort's assistant, Lieutenant A. B. Becher, wrote an article titled "The Log-Book," in which he describes sailors writing on the ship's slate, "this primitive contrivance, one of those clumsy relics of antiquity . . . which should long since have been replaced," haphazard observations of the wind, in which "sprawling characters occupy, with a provoking distinctness, an immensity of space, to the exclusion of some more important remarks." Surprising nobody—surely not the Hydrographer to the Admiralty—Becher suggested replacing that sad state of affairs with "a method for expressing any particulars of the wind and weather, by means of numbers and letters. This method, which originated with Captain Beaufort, . . . is the result of long experience, and affords a concise means of expressing fully the meaning of whole sentences in writing."

It took the Admiralty a while to adopt the scale, though Beaufort didn't drag his own feet. From his position as hydrographer he spread the scale around to anyone who would use it—it "answers perfectly," wrote Admiral George Cockburn from the West Indies in the early 1830s. By 1836 he was already discussing, in a letter to one Lieutenant Baldock, how it could be used by steam vessels (he had no suggestions for how to change the scale's reliance on sail, but he was certain on one point: "pray do not reduce the scale—it is by no means too extensive, and if brought down to 6 points as you propose there will always be a disposition to insert ½ or other fractional part"). In 1837 it was in use by the entire surveying service, and finally, on December 28, 1838, the Lords Commissioners of the Admiralty, "having had under consideration the general utility of recording with clearness and precision, in the Log Books of all Her Majesty's Ships and Vessels of War, the actual State of the Winds and Weather," issued a memo formally adopting Beaufort's wind scale for use. It had no name and was called simply "the annexed scheme," which was of course a copy of Beaufort's scale, with its descriptions of the sail a ship could carry for each force of the wind.

If, with the dawn of modern science all around him, Beaufort made his great contribution through his networking, he still did make a small creative contribution by propagating the wind scale. He sent it on the journey of the *Beagle* with Charles Darwin—which is fitting, since it's an example of how ideas, not just creatures, evolve. The scale began with Tycho, trying to categorize states of wind. With Jurin and Hooke it found expression in a few, but clearer, levels; with the Dutch seafarers and other variations it expanded to more levels, eventually settling on around a dozen. In Smeaton's hands it developed the use of a specific machine as a gauge, and through Dalrymple it was applied—translated, really—to the sea. And with Beaufort's application to sails aboard ship, the scale finally found an ecological niche it could exploit. With the telegraph and other forms of shorthand communication about to change the world (shorthand itself, by the way, was invented in 1839), the Beaufort Scale, like the scientific approach it distilled and the wind it described, was about to spread all over the world.

CHAPTER 6

Getting the Word Out: On the Society for the Diffusion of Useful Knowledge, the Dictionary, and How Sir Francis Beaufort Slept with His Sister

Mad for documentation, Beaufort filled his notebooks with drawings of everything from coastlines to buildings to species of fauna that he encountered.

I WOULD BE TALKING TO FRIENDS or colleagues about Francis Beaufort and the Beaufort Scale, and they would get interested in all kinds of stuff: Beaufort's surprising connections to Darwin and Herschel and the growth of modern science; Beaufort's legacy as a mapmaker and artist; Beaufort's skill in administration and communication; the Beaufort Scale and how Beaufort did and did not write it; and so on. People would listen.

On the other hand, my girlfriend has heard this stuff more than once. So June learned to wait for the briefest pause and then interrupt, saying, "Yes, yes, yes, that's all fine, but did you tell them that *Beaufort slept with his sister?*"

At which point I would have to admit that, as journalists say, I had buried the lead. Yes, Francis Beaufort slept with his sister. Yes, his full, actual, same-mother-and-father sister, and yes, that kind of slept with. And no, it wasn't some sort of easily overlooked childhood experimentation—at the time this all got started, Beaufort was sixty-one, his sister Harriet fifty-seven.

What happened was that in 1834 Beaufort lost Alicia, his beloved wife of more than twenty years, to breast cancer. Beaufort had six children—two older boys, already in school, and a younger boy and three girls, ranging in age from eight to fifteen. Given the times, Beaufort did the natural thing: He called on Louisa and Harriet, his unmarried sisters, for help at home. Harriet, with whom he had always had an especially close relationship, arrived in 1835. On November 26 of that year, Beaufort's pocket diary carries a disturbing entry:

"Fresh hor[r]ors with Har[r]iet, O Lord forgive us."

There are many more mentions, and lots more evidence, though perhaps the less said the better. It went on for about three years, until Beaufort remarried. Then, of course, he had a more appropriate outlet, so the trouble ceased. How Harriet felt about the whole affair only she could say and none should ask; discussions about the options available to an unmarried woman just before the Victorian era I leave to others. But that a man would be so driven to record simply everything that happened to him, that a man would document even something so unsettling—this brings up something about Francis Beaufort that survives still in the Beaufort Scale.

This wasn't a standard part of the biographies of Beaufort that periodically appeared in scientific journals, and it certainly wasn't in Beaufort's obituary. It wasn't known until Friendly wrote *Beaufort of the Admiralty* in 1977, and Friendly only learned it because he did something special.

He cracked the code.

IT WASN'T SO HARD TO SOLVE. In Beaufort's papers at the Huntington Library, on a little slip of lined notebook paper, is a simple code: on the left the letters of the alphabet, on the right a series of squiggles and symbols and altered Greek letters. It doesn't look like it took Friendly a long time to decipher—it's a simple substitution code, one squiggle per letter. Beaufort developed it as a child with his brother.

Throughout Beaufort's early days at sea, his letters home contained passages in code for his brother alone to read—embarrassing things, about women or health issues; when he required help with the truss he wore for the hernia that almost all men at sea developed as a result of the constant lifting and pulling, for example, he wrote in code. But the code never left him, and as his life went on he continued to use it even in his diary, when he wrote things especially private—like, for example, that he was sleeping with his sister.

Beaufort's relationship with his sister is not especially important in itself—a sad and embarrassing chapter in a marvelous life; even his biographers wrestled with whether to include it. But the journal entries show not only Beaufort's commitment to documentation but also his fascination with communication, with finding ways to express complex ideas simply. The wind scale, after all, is a kind of simple code—a wind that requires you to double-reef your topsails

P 296

R to L	O Smyrna what I have suffered within
L to R	your wals (CQ)! The beastly lasciviousness and acute cuming of
R to L	one woman, they seductive charms, they dangerous
L to R	folies of another and the strange mixture of my own
R to L	vices, virtues and weaknesses have thrown my mind
R to L	and body into a state whence they must
L to R	have long to emerge. Oh God let me turn
R to L	this biter [illegible] (CQ) a leson (CQ) to acount (CQ):: reclaim the
L to R	one, and forgive the other, [illegible] heal me of God

Beaufort's love of transfigured information—and almost obsessive need for documentation—both show up in this journal entry and its translation.

and reef your jib is reduced to "moderate gale," or even, more simply, 7. He did the same thing with weather too, by the way—again taking a system designed by Dalrymple and included in the book Dalrymple gave to him, Beaufort instituted a simple weather code in his logbooks. For example, *b* would mean blue sky, *c* clouds, *s* snow, and *q* squalls, enabling four or five letters in the log to express a good paragraph of information.

This was the type of thing that fascinated Beaufort, who always strove for simplicity, clarity, and distillation in communication. The scale takes a lot of information and reduces it to a single numeral; it's no surprise to learn that in his war service Beaufort served often on what were called repeater frigates—ships whose job was to observe the signal flags displayed by the commander and repeat the signal up and down the line. And though Beaufort is known as a friend to science rather than as a scientist, he did make one major attempt to develop something scientific. It involved coded communication.

Beaufort had a lifelong friend and mentor in Richard Lovell Edgeworth, a member of the Birmingham Lunar Society and originally a friend of Beaufort's father. (The relationship was always happy and grew complex: Edgeworth eventually took Beaufort's sister for his fourth wife; Beaufort took Edgeworth's daughter Honora as his second.) Among Edgeworth's pet projects was what was then called a telegraph. First developed in France in 1793 by Claude Chappé and called a semaphor, the telegraph in the eighteenth century was basically an application of ships' signal flags to the ground, except instead of flags of specific colors or patterns to convey meaning, the telegraph used pointers, visible from afar. Different positions signified different numbers, and numbers corresponded to words in a coded dictionary. With stations ten miles or more apart (operators were equipped with telescopes), messages could travel from, for example, Lille to Paris (the route of the first line), a distance of about 120 miles, in less than an hour.

People had pursued simple systems of long-distance communication since Aeneas supposedly sent messages using torches, but Chappé's was the first that worked predictably and regularly; lines of those mechanical telegraph stations eventually spanned France and other European countries.

Edgeworth—with help from Beaufort—went to work on something similar in Ireland in the early 1800s, after Beaufort had returned from his participation in Nelson's fleet and before he was assigned to the *Woolwich.* Edgeworth proposed a telegraph line from Dublin to Galway, using a system he had designed himself. The idea was great (and Edgeworth may have had it as early as 1766), and it certainly fit well with Beaufort's lifelong efforts to, as the 1832 article about the Beaufort Scale in the *Nautical Magazine* put it, find "a concise means of expressing fully the meaning of whole sentences in writing." In fact, perhaps too much so: Beaufort and Edgeworth created a system so preposterously complex that its code book measured four feet by two feet. Still, they improved the system somewhat and by 1803 had convinced the Irish government to pay for the line. Beaufort managed the construction of thirty stations and the training of thirty teams of peasants to operate the complex signaling machinery and codes.

Regrettably, whether because of poorly educated operators, misty Irish weather, or the general complexity of the endeavor, the "tellograph," as Edgeworth called it, was a complete failure. A July 1804 demonstration in front of the Lord Lieutenant of Ireland was a disaster, and after a late-July test, Beaufort wrote to Edgeworth that "not one word of the whole [came] out right." What Francis Beaufort had to contribute to science was through written, not mechanical, communication. The project was abandoned.

BUT BEAUFORT'S INTEREST IN REDUCING LANGUAGE to an easily transmissible format never diminished; that is exactly the point of the wind scale, which Beaufort wrote in his journal in 1806—less than two years after the failed tellograph—and used privately for the next three decades. Thus its adoption in 1838 for official use by

Admiralty vessels must have been a great satisfaction to him—though oddly his journals and letters never mention it. But as the scale's development and spread over the subsequent decades demonstrates, Beaufort loved to distill language not just into code. He also loved merely to express ideas in common language—in the simple language that even after his death characterized the wind scale, especially as it eventually appeared in the dictionary.

This was the early scientific era, after all. You might not just say, "Look! A cloud!" anymore—after 1804 you'd say, "Why, it's a cumulonimbus." Before, there was air—now there was a complex mixture of nitrogen and oxygen. The Linnaean classification system was accurate and marvelous—as long as you knew Latin. Beaufort, in this time of rapidly specializing language, opted always for the simple, the mundane, the common. We need to know how windy it is, but for Beaufort that doesn't mean we need a whole raft of new technical terms. Good ordinary words will do. Beaufort applied the clarity and simplicity he used on his charts and maps to general description as well.

As a result, the development of the wind scale itself becomes a kind of measure, as the common words it uses give detail about the times in which it appeared as, after 1838, its use spread.

In the 1840s, Samuel Morse's improvements made the electric telegraph the explosive communications advance that Beaufort and Edgeworth's mechanical tellograph had failed to be. The first long-distance telegraph—Morse's famous "What hath god wrought!"—traveled from Baltimore to Washington on 1844; thereafter telegraph lines crisscrossed Britain, North America, and Europe almost immediately. A line crossed the English Channel by 1850, the Atlantic by 1866.

Among the earliest information sent back and forth was weather data. Almost immediately after the telegraph began to spread, Joseph Henry, the first secretary of the Smithsonian Institution, made a trade: he would provide the meteorological instruments to telegraph companies if they would send readings by telegraph to the Institution daily. By 1853 Henry's network had thirty-two stations (by 1860 it numbered more than a hundred), and so much weather communication was taking place that Matthew Fontaine Maury, the great American hydrographer (he is often called Beaufort's American equivalent), organized the First International Meteorological Conference, in Brussels. The Beaufort Scale, with its widely understood categories and by then entrenched usage among Admiralty sailors, was well suited for common use; and with its reduction to simple numerals, it was perfectly suited to telegraphic transmission. At that August 1853 conference, the Beaufort Scale was adopted as the standard for international communication of the force of the wind.

By 1862 Beaufort's scale had been adopted by the British Board of Trade; in 1874, changes in sail use (instead of a single topsail, most ships now had upper and lower topsails) required small updates, but the scale continued to measure the wind, at sea and on land, in a growing number of places throughout the world.

However, just because a tool "answers perfectly," as Admiral Cockburn said of Beaufort's scale, and just because it does its job with linguistic perfection is no reason to expect that its passage from interesting idea to international standard will progress without detour. The path of the Beaufort Scale from Smeaton's windmill to the Merriam-Webster Ninth *New Collegiate* contained a few more hurdles.

FOR ONE THING, INTERNATIONAL COOPERATION generally takes its time; just because some group of meteorologists calls something a standard doesn't make it so. Many other wind scales were in use. The descendants of the scales developed by the Scandinavian mariners were still being passed around among sailors, and other variations appeared as well. In fact, by 1896, when "An Attempt to

Determine the Velocity Equivalents of Wind Forces Estimated by Beaufort's Scale" was published in the *Quarterly Journal of the Royal Meteorological Society,* its author listed twenty other wind scales "in which the different winds are described as words," all published after 1844.

Some were plainly nothing more than variations on the Beaufort Scale. Other scales reach further back—a scale from an 1845 meteorology book is nothing more than the 0–5 scale Jurin had suggested in 1723, and one from 1849 is an exact copy of the scale Smeaton made in 1759, before he even found the windmill to base it on.

The problem the wind scale was designed to solve was exacerbated, of course, by the multitude of scales. Two six-point scales, both from 1863, agree on the following progression of winds: "moderate; fresh; strong; heavy; violent." But one starts with "light" and the progression follows; the other starts with "moderate," runs through the progression, and ends with "tremendous." One correlates "strong" with 7 or 8 on the Beaufort scale, the other with 5 or 6. Like the man with two watches who is less certain of the time than the man with none, observers faced with the profusion of scales could actually find themselves more confused about how to describe the wind than they had been before the scale came along in the first place.

JUST THE SAME, THE SPREAD OF WIND SCALES would have delighted Francis Beaufort. The scales were showing up in handbooks of meteorology, in technical dictionaries, in magazines, journals, board of trade circulars. I eventually stumbled on the Beaufort Scale in a dictionary, and reference works like that would have pleased Beaufort in themselves. The mid-nineteenth century was, in fact, a great age of reference books, the moment when such books, as we would recognize them now, came into being. Modern almanacs had first appeared around 1700, though they commonly included a lot more astrology than highly uncertain weather predictions; only as astronomy advanced through the 1700s did almanacs become exact and useful—the Royal Observatory did not publish the first volume of its *Nautical Almanac,* whose astronomical listings were accurate enough to be used for navigation, until 1766 (the Scientific Branch of the Admiralty took that function over in 1831). The first recognizably modern encyclopedia—a compendium of learning on every subject, organized alphabetically and cross-referenced—was Ephraim Chambers's *Cyclopaedia,* published in Britain in 1728; previous encyclopedias had been organized according to complex systems dividing knowledge into separate areas, making them far more useful to scholars than to the general public. But as the middle class grew and more people became educated enough to want information at home and wealthy enough to afford it, new sources appeared.

The French grew interested in the Chambers *Cyclopaedia* and asked for a translation. Chambers declined, so the French used his work as the basis for their own work, the famous *Encyclopédie,* edited by Denis Diderot and Jean d'Alembert and considered the ultimate expression of the Age of Reason. By 1771 the British had again responded, with the first edition (in three volumes) of the *Encyclopaedia Britannica.* Samuel Johnson published his dictionary in 1755; Noah Webster did the same in the United States in 1828, and the *Encyclopaedia Americana* followed in 1829. In fact encyclopedias were everywhere in the first half of the nineteenth century, dozens appearing, and usually failing to compete and then dying.

These kinds of general compendia were, quite literally, near to the heart of Francis Beaufort. The tiny, leatherbound pocket diaries he carried with him were called "The Gentleman's Pocket Memorandum Book," and they included little things like lists of different varieties of British knights (of the Garter, of the Thistle, of the Bath); of the House of Peers, the directors of the East India Company, the baronets of Scotland. On a more practical note they provided a guide to hackney coach fares (you could go half an hour for a shilling), assessed taxes on such things as hair powder, dogs (fourteen shillings for most, but a pound for a greyhound), horses, servants, and windows. There are lists of mail coaches and what time they go through different towns; country bankers; bank holidays; stamp duties; and the tables of weights and measures and monthly almanacs in millions of typefaces exactly like we would find in the *Farmer's Almanac* today. In one sweet indication of Beaufort's character, inside the cover of his 1848 journal he's written in flowery script, "KCB." In 1849 he doesn't need to, because he appears in the list of Knights Commanders of the Bath, the honor he received, he wrote in his journal, from "the august (but plump) little hand" of Queen Victoria herself. Along with his belated advance to rear admiral in 1846, the honor, at age seventy-five, finally gave him the recognition he had craved all his life (though in order to continue as hydrographer he had to retire officially from active service, thus, predictably unsatisfied, becoming what was called a "yellow admiral").

More than just those pocket diaries—and the scale he left behind him—demonstrate Beaufort's love of lists, these compendia of learning. Early in his days at sea, he regularly wrote to his father begging for books; by the time he sailed on the *Woolwich* he carried with him a library of 202 volumes. The books he requested from his father were "a general and moral treatise on all the parts of Philosophy. . . . Secondly I want one which treats of all the Different Sciences in a very particular way. Such as Natural Phil: Mechanics, &c. &c. . . . And lastly—I want one entirely devoted to Astronomy which treats it in a very scientific way both theoretical and practical." By the next year he sent his father an inventory showing he had already accumulated, in addition to his nautical books, some sixteen titles, including a ten-volume set of Shakespeare, his father's memoir and map of Ireland, general references on astronomy, the natural sciences, and arithmetic, as well as, of course, the Bible and an early dictionary.

BUT BEAUFORT'S LOVE OF REFERENCE BOOKS manifested itself most profoundly in another way. In 1826, before he became Hydrographer to the Admiralty, Beaufort joined a group of gentlemen founding the Society for the Diffusion of Useful Knowledge, a utilitarian group whose 1829 prospectus listed as its goal "imparting useful information to all classes of the community, particularly to such as are unable to avail themselves of experienced teachers, or may prefer learning by themselves."

The society's projects included maps that subscribers could collect to make an atlas; the *Penny Cyclopaedia;* a series of books called the Library of Useful Knowledge; a *Biographical Dictionary;* and dozens of other edifying publications produced singly or in series (the Library of Entertaining Knowledge, for example, was a series of popular books on geography, biography, and other such subjects). Beaufort, in the 1820s looking for something to do, became the Society's map editor, a job he kept long after he became Admiralty Hydrographer in 1829. He famously arose early in the morning and spent several hours on Society maps before going to work at the Admiralty and doing the same thing there. Because the goal was to sell the maps as cheaply as possible, Beaufort refused payment for his Society work, hoping thereby to get the maps into more homes. He even urged, in a letter to the promoter of the map project, that the Society not limit itself to its large atlas: "might we not attempt still higher game?" he asked. "Might we not produce a [smaller] Atlas . . . [and] would not a Copy find its way into every house in the Empire & would not this be diffusing real tangible knowledge?"

Regrettably, the Society fell victim to its own high goals. For one thing, producing so many different types of publication reduced the

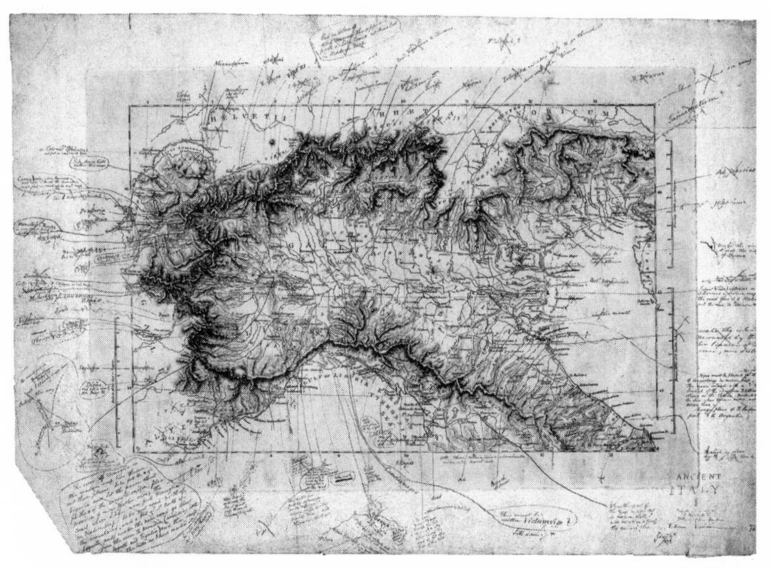

Beaufort's almost maddening eye for detail slowed production of maps by the Society for the Diffusion of Useful Knowledge. This map demonstrates how.
This item is reproduced by permission of The Huntington Library, San Marino, California.

attention paid to each. For another, publications came out much more slowly than the Society had planned, frustrating subscribers. Especially on Beaufort's maps, that famous attention to detail eventually became a problem.

But in that attention to detail lies Beaufort's spirit. At the Huntington Library, among his other papers, are several maps Beaufort worked on for the Society, each covered in hundreds of the tiniest, most careful changes imaginable. He puts in the absent names of tiny rivers, catches minuscule drawing mistakes, crams in further information (on a road through the Alps he's suggested adding "Hannibal's route in very small print in direction of the road"; on a road through Beloochistan (Baluchistan) labeled "thirteen days journey for laden Camels" he's added "mountainous road"—presumably so a reader wouldn't have to wonder what was taking those camels so long). Beaufort, so painstakingly detailed, was clogging up the works. "I hope you will be able to bring Capt. Beaufort's mind to the conviction that a little imperfection now and then, in the Maps . . . is better than the present slow progress—the ill effects of which we feel daily," the map publisher wrote to the Society's secretary in 1836. It never got much better, and by 1848 the Society was gone.

It did have its successes—more than 3 million maps were printed and sold, and tens of thousands of other books. And as the Society wrote about itself in its own *Penny Cyclopaedia,* "besides the direct benefit conferred on the public by placing in their hands works carefully prepared, vast advantage has been derived from the example of the Society's success." For example, the *Penny Magazine* was the inspiration for many other magazines dedicated to the education of those for whom "the greatest difficulty . . . has been the obtaining of books." Society books were translated into other languages, and the 1842 *Penny Cyclopaedia* itself remains a valuable source of information on what Victorian society was trying to teach itself. And the Everyman's Library and the Harvard Classics and every publishing series meant to fill the shelves of average people with everything they need to know owes a small debt of gratitude to the Society, and to Francis Beaufort.

AMONG OTHER WAYS, INFORMATION spread through Beaufort's beloved maps, and one vital piece of information the maps spread was the names of the places represented—a genuinely important issue for Beaufort and the scientists of his time. European people—especially the British—were coming into close contact with places all over the globe, and they were drawing the first usable maps, and the names they wrote down might affect how places thereon were known for decades, even centuries. Anyone who's ever been to Quincy, Massachusetts, or Reading, Pennsylvania, (KWIN-zee and REDD-ing, if you're wondering), knows how the indigenes respond to visitors who don't even bother getting the name right. For one approach to this subject, consider James Capper, whose name has come up here before; his 1801 treatise on monsoons and winds included an early reproduction of the scale that John Smeaton developed.

Capper believed that the more people knew about the new places that through exploration were becoming familiar, the more quickly he and others like him could perform such vital services as fixing their names. According to Capper, place names should be above all geographically accurate. He gives an example as he describes one of his charts: "In the map therefore I have adopted the name of the Arabian Gulf, for the Red Sea." He liked gulfs—he did the same east of India, putting a gulf in place of "what is usually termed the Bay of Bengal." No matter that half the world spoke of the Red Sea; who cared what the inhabitants called it. Let's call a gulf a gulf, and never mind the locals.

To us now this looks like a silly species of early-nineteenth-century arrogance, which of course it is, but it's important to remember that this was one of the issues of the day.

Beaufort knew this. In fact, this knowledge distinguished him. When he made his voyage to southern Turkey, surveying land not thoroughly investigated by Europeans since the ancients, he consulted his Strabo and other classical geographers. But he also heard what the locals called a place, and he considered how other foreign visitors referred to it. At one point he writes about a place called "Kastelorizo." Since it once held a fortification jutting from between cliffs with a reddish tinge, he judged it might have picked up its name from the Italian "Castelrosso . . . for we find that many sea terms, as well as names of places, have been adopted from European sailors."

Big deal. Kastelorizo doesn't mean anything in Turkish or Greek, but Beaufort is not about to come up with an English or Turkish name that makes more sense to him. What it's called is what goes on the map: "It is now called and written as above," he says. "And it appeared to me more judicious to retain the vernacular names, wherever they could be distinctly ascertained, than to adopt those applied by other foreigners." As for thinking up a new name to suit himself, "The custom of inventing new names is still more pernicious to the true interests of geography."

It's beautiful—it's just like the *Manual of Scientific Enquiry.* "The true interests of geography"—which are, plainly, to describe what you see and get the information out there in the most useful manner possible. Beaufort's certainty here is inspirational. In a time when how to present information on maps was a highly imperfect art, he knew that as a sailor and a traveler, he needed the *real* information, not the impressions that someone new to the area thinks are logical. A map drawn by someone like Capper could prove desperately dangerous to a ship. Imagine a ship planning to stop for fresh water near a stream the locals called Red River. If the cartographer who created the map had decided that since the river came from the mountains it really ought to be called the Cold River, and labeled it so, the ship could miss its fresh water and end up dry, off course, or worse. The name of a place is part of the place—an attribute, just like its height or its color. Beaufort wanted to accurately represent what he saw, and his maps show it.

It's almost being mean to point out that we refer to the Red Sea and not the Arabian Gulf, so Beaufort and those who thought like him obviously won the case, with Capper sailing upriver into a cartographic evolutionary dead end. But it's a wonderful thing to think about, especially when we have the luxury of printed maps of every scale. It's worth thinking about what would have happened if men like Beaufort had not stepped to the fore. In 1999 the Mars Climate Orbiter was lost because engineers forgot to convert miles to kilometers. Had mapmakers fallen into the habit of constant renaming, similar problems of translation would have plagued ships. Incorrect information is dangerous, and Beaufort knew it.

In fact, Beaufort's zeal to be sure he knew not only the accurate names for what he saw, but to get as much accurate information as he could from locals, became legendary, and it still affects maps and those who read them, both by what is and what is not on the maps.

For one thing, as Hydrographer to the Admiralty, Beaufort for twenty-five years presided over the creation of the best and newest maps in the world. Surveyors had a natural tendency to use new places on the maps they created to curry favor—think of the number of Victorias or Charlestons in this world, to say nothing of the places named after purse-string-holding functionaries whose identities have survived only on the maps their subordinates littered with their names, hoping for favor and advancement. The charts generated under Beaufort were the opposite. Because of his respect for natural names—and perhaps because of the struggles he endured himself trying to advance in a system that rewarded influence faster than it rewarded excellence—Beaufort even removed his own name when it showed up in a newly charted bay or atop an island.

That's not to say his name is lost, though. For one thing, when the men searching for the Northwest Passage put his name on a sea north of Alaska, he left it there, perhaps because so much of the final years of his life involved work with the committees concerned with seeking that passage—and seeking the explorers lost in that endeavor. Even more important, though—and leaving the Beaufort Scale out of it entirely—his name remains a more powerful legacy. Surveyors in the service of Beaufort well knew of his great desire for detail and accuracy. They thus commonly hired native guides so that they would get things right—and if they found a guide who was truly excellent, who knew where everything was and what it was, who helped them pilot the boat and take bearings and soundings and gather crucial information, the surveyors would jokingly call him Beaufort. It took, and to this day there are parts of China and Borneo where the name Beaufort persists among the descendants of those guides. That's a lovely irony for someone too diffident to leave his name on a map. Those Beauforts, a type of intellectual offspring, are part of the heritage—in this case mere names, a *linguistic* heritage—left behind by Sir Francis.

WITH THE END OF THE SOCIETY for the Diffusion of Useful Knowledge, Beaufort focused only on his Admiralty work, and it's no surprise that the *Manual for Scientific Enquiry* appeared in 1849, the year after the Society met its demise. It was in spirit a Society book, designed for "Travellers in General" and meant to edify and train the observer, who was expected to be not a scholar but merely a man of "good intelligence and fair acquirement."

But the *Manual* was one of Beaufort's final great contributions. At the time of its publication he was seventy-four, and though he was slyly mentioned in the section on hydrography as "the able and indefatigable officer at the head of the Hydrographic department," he was growing somewhat less able and, finally, for the first time in his life, fatigable. As what amounted to the world's head nautical surveyor, he spent much of his final years in the Admiralty as a member of the Arctic Council, a group of men focused on the furious search for the Northwest Passage—and, of course, that group spent a great deal of its time organizing missions to search for those ill-fated explorers who had initially gone to search for the passage. The search for the Northwest Passage was in some ways the minor-key coda to the end of the age of sea exploration. The great missing southern continent had turned out to be Australia and Antarctica, and the oceans had been traversed from every direction. The Northwest Passage was, in a way, the only undiscovered country left for a generation of men who had grown up sailing boats where no boats had gone before. That probably explains why even when it became quite plain that the Northwest Passage, if it even existed, would lie much too far north to be of any value, the search never slowed. Used to the heady excitement of drawing coastlines never before mapped, Beaufort and his cohort can be forgiven for refusing to give up.

Beaufort's role in this British obsession has an unusual connection to the United States, by the way. In 1845 Sir John Franklin, a personal friend of Beaufort's, left England with the ships *Erebus* and *Terror,* in search of the Northwest Passage, with three years' worth of supplies. Those ships were soon crushed in the Arctic ice and abandoned by their crews, who ultimately died trekking overland. When no word had returned by 1848, rescue missions were mounted, and over the next years several also unsuccessful search parties were sent, by sea and land, from east and west. One of those ships, the *Resolute,* saved the crew of one of the previous search parties, though the *Resolute,* too, eventually lodged in the ice and was abandoned. Embarrassingly, it thereafter broke free and drifted east, where it was salvaged by an American whaling ship and eventually presented back to the Admiralty as a gift. The ship was subsequently dismantled, and a desk made from pieces of the *Resolute* was presented in 1880 by Queen Victoria to President Rutherford B. Hayes. President Kennedy brought it into the Oval Office, where it's been on and off since then. Today President Bush sits behind a desk made from a ship sent on its greatest adventure by Francis Beaufort. As for that mission, Eskimos finally relayed the news of the death of Franklin and his crew to a geographical expedition in late 1854, though full details were not known until 1859, after Beaufort's death.

Beaufort was ill in his later years, and if there was much keeping him happy at the Admiralty, Franklin's death probably put an end to it. Beaufort attempted to retire in 1854, but with the Crimean War impending, the Lords of the Admiralty persuaded him—at age eighty—to remain at his post to help create charts needed for the war. He did so for another year, in failing health, until at last in 1855 he resigned for good; a portrait of him was commissioned and hung in the Great Hall of Greenwich Observatory. Finally at home, Beaufort continued his weather observations until the end. By then, his pocket diaries show, with nothing to observe in his last years of illness, he was even documenting his bathroom habits. That's a little sad but also a little beautiful: the compulsive record-keeper who paid attention to everything, reduced to his final observations. With family near, Beaufort died in his sleep in December 1857.

BUT OF COURSE THE WIND SCALE KEPT ON—it could not be otherwise. Beaufort's obituary makes no mention of it at all; in the 1859 third edition of the *Manual for Scientific Enquiry,* the scale is known only as "Admiral Beaufort's system of abbreviations." But by 1874 the first article trying to resolve different wind scales appeared in the *Quarterly Journal of the Royal Meteorological Society,* titled "A Relation between the Velocity of the Wind and its Force (Beaufort Scale)." Beaufort's name had finally and forever been tied to the wind scale.

With the continuing expansion of telegraphic weather communication, the several competing wind scales throughout the world conflicted more and more, but in the comparisons of different scales that appeared here and there, the Beaufort Scale was always included, always the scale against which the others were compared. And when in 1906 the Meteorological Office in London, finally addressing the obvious fact that a scale based on the sails of ships at sea was not helpful for observations on land—and with the advent of steam was no longer even particularly helpful for ships at sea—decided to create an updated scale based on coastal and shore observations, the title of the publication was "The Beaufort Scale." Beaufort's name on the wind scale was already becoming his legacy.

And the scale was appearing in more than just meteorological journals. That 1906 scale was the one with the brilliant and lucid descriptions, the one I encountered eighty years later in the Merriam-Webster dictionary. Its journey there, from the Met Office, was fairly straightforward. As meteorological knowledge increased, nation after nation created national weather bureaus—the 1853 International Meteorological Conference in Brussels had recommended as much. In England the new meteorological department was part of the Board of Trade, and at its head was Robert Fitzroy, onetime captain of the *Beagle.* Based on a rather small telegraphic network, Fitzroy began gathering data on weather and ocean currents (he made the information available to sea captains in a coffee shop they frequented—it was called Lloyd's of London). In reaction to a catastrophic 1859 shipwreck, in 1861 Fitzroy began not merely making data available but issuing storm warnings and rudimentary forecasts. Beyond sea captains, those were made available for, and printed in, newspapers.

The state of meteorology was so simple, however, that Fitzroy's forecasts were commonly wrong, and he was derided in the press for their inaccuracy. The ridicule hit him hard, and he killed himself with a straight razor in 1865. Things went a little bit better in the United States, where in 1846 Matthew Fontaine Maury had begun publishing charts that displayed marine temperature, current, and wind. The telegraphic network started in 1849 by Joseph Henry of the Smithsonian proved far more accurate than Fitzroy's smaller network. For one thing, Fitzroy's telegraphic stations were largely in the British Isles and on the Continent, giving him little foreknowledge of the weather over the sea, from whence most of his storms arrived; Henry's stations, on the other hand, stretched westward across a continent, documenting weather systems generally moving eastward; to describe storms approaching the eastern seaboard, where the bulk of the population lived, Henry had all the advantages. By 1853 weather charts were being printed in the New York daily newspapers. And once you start putting weather charts in the newspapers, new words start floating around. The writers of dictionaries, looking to document the language of their populations, begin to prick up their ears.

In 1909, *Webster's New International Dictionary,* under "wind scale," included a table of several different scales; the Beaufort Scale appeared at each end, as the standard. In 1934, in the *Second New International,* the entry under "wind scale" remains, but there's also an entry for "Beaufort's Scale," though it refers you to the "wind scale" entry.

Finally in 1961, with the *Third New International,* the "wind scale" entry remains, but without a chart, and it refers you to the entry for "Beaufort Scale," where you find the lovely passages of the 1906 scale, finally in the most important dictionary in the United States (not especially quickly, given that it was officially adopted by the International Meteorological Organization in 1939). It had been included in the *New Collegiate Dictionary* (the desktop version of the *New International*) starting with the sixth edition, in 1949. It's been there ever since, waiting for me until, in the Ninth *New Collegiate,* I found it.

FOR A LONG TIME AFTER I FIRST DISCOVERED the Beaufort Scale in the dictionary, whenever I was in a used-book store I would make a beeline for the reference section and look it up in every dictionary and encyclopedia I could find—some old white-pebbled, two-volume Funk & Wagnalls, or a Larousse, or any number of other old dictionaries that didn't make the evolutionary cut and ended up on used-book shelves.

Partially I looked for the scale to see at what point it had made its way into general usage; and partly I checked to see in how pure a form it appeared. Was it accurately copied? Was it complete? Surprisingly often, it was neither. I found scales with the wrong wind velocities and scales with the wrong wind names, scales based on some of the non-Beaufortean ancestors I eventually learned about. Beaufort scales that stopped at 10, Beaufort scales that rocketed to 18. A surprisingly common error had Beaufort writing the scale in 1805, not 1806. It was kind of pointless, but I began to take pride in knowing when the reference works were wrong. But I was looking for something more in those sources, too. I had discovered the Beaufort Scale in a dictionary and fallen in love with its language, then set out to learn what I could of its author, to see whether he had left more great writing for me to read. It all turned out to be much more complex than I had expected, leading through sailing and engineering and science and technology. After finding myself pretty far afield, returning to dictionaries felt like coming home.

And this dictionary lust didn't affect just me. Once my fascination with the Beaufort Scale had surpassed mere interest and I had accumulated a thickening file of strange Beaufort Scale photocopies, letters, and artifacts, I could be depended on to pull it out to entertain friends who came to dinner. I worried I had become like one of those people who are obsessed with the overlarge electric train setup in the basement, or like the old woman in the joke with the trunk full of pancakes in her attic—that my fascination wasn't dangerous, but perhaps I should just keep it to myself. Instead, people expressed interest, and one of the things they liked to do was go to the Merriam-Webster Ninth *New Collegiate* and scan the spread of pages the scale fell on for its companion words. "Oh, look!" one friend cried. "You have 'be-all and end-all'"! Also "beautiful people" (with or without initial caps), "Beau Brummell," and, especially delightful because of its proximity to the wind scale, "becalm."

This is hardly exceptional behavior, especially among editors. An editor I once worked with had been such a lonely kid that he spent his time reading the dictionary; he entertained us at work by challenging us to cover the words on any page of the dictionary and show him only the animal sketches. "Gyrfalcon," he would say without hesitation, or "manatee—come on, give me something hard." A certain amount of geeky wordishness comes with the territory, but I believe there's more to this dictionary fetish than the idle interest in words and wordplay shared by most readers, writers, and editors. I think love of dictionaries is in itself a species of observation, of understanding the world through its abstraction. I think a dictionary is a sort of Beaufort Scale of words. Just as the Beaufort Scale is a shorthand, a code for wind behavior, the dictionary performs the same function for an entire language.

For one thing, the dictionary is an example of organization every bit as beautiful as the Beaufort Scale—though entirely differently conceived. The organization according to first letter is as exactly appropriate a tool for organizing words as organization by force is for the winds. The winds organized alphabetically would simply make no sense; and the words organized according to force—word results, I guess—would be something closer to a grammar than a dictionary. It might be interesting and it would probably be fun, but it would only be good for some specific purpose, if even that. No, the dictionary and the Beaufort Scale are organized just perfectly for their diverse tasks.

BUT THERE'S MORE THAN PURE ADMIRATION in my incessant looking up of the Beaufort Scale. In its form, and in what surrounds it, it often sheds light on the reference book—or on the time in which that reference book appeared.

Here's an example. I once bought in England an early pocket

dictionary—it's about the size of a mystery paperback you'd buy today, though it's bound in tooled brown leather. From its inscription I'd say that one Henry Jones received it as a gift on January 31, 1876, and that Mr. Jones had lovely handwriting. (I also guess that a subsequent owner received it in 1918, since a little subtraction problem penciled in the preface margin yields the forty-two-year difference. I imagine the owner delighting in owning something so old.) All fun enough, and when I bought it in 1980 I did my own subtraction and congratulated myself on buying a hundred-year-old book for twenty pence.

But my favorite thing in that dictionary is the way the words and pictures it contains fix it in time. There can't be many dictionaries before it, for example, that define—and even picture—a locomotive, which in this case looks like a simple drawing of a choo-choo, because that's what trains still looked like (though they weren't yet called trains, or at least there's no such definition for "train" in this dictionary); and there can't have been many after it that would have expected its readers to need the definition for "Belial," one of the names for Satan.

But what I return to again and again in the 1876 *Illustrated Pronouncing Pocket Dictionary of the English Language on the basis of Webster, Worcester, Walker, Johnson, etc.* is the definition of "goat." There's a drawing, of course—actually it's a great page for drawing, also including pictures of grapes, Gothic arches, a gong, and a gorilla. The goat stares placidly ahead, his curling horns identifying him more than his shaggy coat. The book describes him thus: "a ruminating animal, seemingly between a deer and a sheep."

I love that definition for two reasons. For one, it organizes animals without much discussion or fuss. The dictionary tells you it's a ruminant, but it figures you'll already know the basics—mammal, four legs, hooves, horns. And the definition itself implies a categorization strategy, an order. Whatever the order of four-legged ruminating mammals with hooves, if you go from sheep to deer, you're going to have to go through goat to get there. You're getting larger, you're getting slimmer, you're getting bigger horns. It's a recognizable progression. Go the other direction from sheep and you'll probably end up at pig; keep on going after deer and I suppose you head toward elk or even reindeer. The point is, it's a continuum. Not necessarily a scientific continuum—nobody is saying a thing about the relationship between the species of sheep, deer, and goat. All the dictionary is trying to do is to make sure that if you read about one, you'll know what it is. "Oh, yeah," you might think. "I saw one of those out at a farm—it was too big for a sheep and too small for a deer. I was wondering . . ." And now you know.

It's beautiful—it's a definition that does exactly what it must, in ten words and a tiny picture smaller than a dime. It does its job.

The second reason I love this definition is what it tells you about its readers, which I guess is really an extension of the first: This is a definition of a goat *for people who regularly saw goats.* This dictionary was printed in Glasgow in 1876, and if you lived around Glasgow during that time, then in your peregrinations in an average week, you'd run across some goats—in farms and fields, in markets, goats were a part of your life. So all the definition needed to do was give you a running start on where that goat fit into the greater scheme of things and it was on to the next (in this case "goatish," which means not smelling very nice).

A dictionary thus becomes a document, a living history, a portal into the world it means to describe. The Merriam-Webster 1934 *Second New International,* for example, includes color plates of the house flags of the major steamship lines—it's a detail, a clue about what was important in 1934. I once bought an atlas printed in 1933 only because in every map showing the North Atlantic it included transatlantic cable lines; in maps of Europe it showed the tangle of cables running all over the North Sea and the Mediterranean. That's what was important then, and it's sweet to remember it now, when it would no more cross the mind of an atlas publisher to include uncountable transatlantic cables than it would to include mail routes. The atlas, a reference book, itself becomes an artifact instead of merely a guide to others.

AND THE BEAUFORT SCALE DOES EXACTLY THE SAME. I first learned this when in 1993 I got the Merriam-Webster *Collegiate Dictionary,* Tenth Edition—the next edition after my prized Ninth. Surprising nobody, I immediately opened the book to the Beaufort Scale.

And I found it changed.

Not hugely—it wasn't suddenly translated into the decimal system or wholly rewritten. Still, there were changes.

Some of them looked like the result of well-meaning attempts to save words—in force 2, "light breeze" has gone from "ordinary vane moved by wind" to "wind vane moves." That saves two words, though I'd have filed it under "if it ain't broke" myself. Force 5, though, goes from "small trees in leaf begin to sway" to "small-leaved trees begin to sway." That's a plain change in meaning—a small tree in leaf is one thing, a tree with small leaves is another, just as a small man with wit is not the same as a man with small wit. In seven, "whole trees in motion; inconvenience in walking against wind" has changed to "whole trees sway; walking against wind is difficult," and without becoming too pedantic about the difference between being "in motion" and "swaying," I can say we've lost a good deal of the inchoate loveliness of the original. Well enough has manifestly not been left alone.

In force 6, the wires still whistle, but what in 1982 were still charmingly called telegraph wires have now become overhead

ESTIAL 1 2
; n
avy loads or
; [ME *beten,*
1 : to strike
en used with
orcefully and
usly **e :** to
as if in quest
n *up* **g :** to
a drum⟩ **2**
er, paste, or
d (1) : to
peated strik-
esp : to flat-
beat **3 :** to
T; *also* : SUR-
he odds⟩ **c**
leave dispir-
(1) : to act
n in advance
: system⟩ **d**
e against (a
vi **1 a :** to
with oppres-
at a drum **2**
ng struck **3**
c : to strike
r or as if for
windward by
: **about the**
: to the point
at it 1 : to
ins out : to
— **beat the**
— **beat the**
olicize vigor-
es connected
physically or

s; *also* : PUL-
c : a driving
ce of a time-
ie ~⟩ **b :** a
a : a metri-
ffect of these
musical per-
ristic driving
t excels ⟨I've
head of com-
ard **b :** one
ons of ampli-
electric cur-
(as of one leg
lj
: being in a
or morale **2**

ed shape ⟨~
R ⟨a ~ path⟩

(1920) : DANDY 1
beau·coup \'bō-(,)kü\ *adj* [F] (1918) *slang* : great in quantity or amount : MANY, MUCH ⟨spent ~ dollars⟩
Beau·fort scale \'bō-fərt-\ *n* [Sir Francis *Beaufort*] (1858) : a scale in which the force of the wind is indicated by numbers from 0 to 12

BEAUFORT SCALE

BEAUFORT NUMBER	NAME	WIND SPEED MPH	WIND SPEED KPH	DESCRIPTION
0	calm	<1	<1	calm; smoke rises vertically
1	light air	1-3	1-5	direction of wind shown by smoke but not by wind vanes
2	light breeze	4-7	6-11	wind felt on face; leaves rustle; wind vane moves
3	gentle breeze	8-12	12-19	leaves and small twigs in constant motion; wind extends light flag
4	moderate breeze	13-18	20-28	wind raises dust and loose paper; small branches move
5	fresh breeze	19-24	29-38	small-leaved trees begin to sway; crested wavelets form on inland waters
6	strong breeze	25-31	39-49	large branches move; overhead wires whistle umbrellas difficult to control
7	moderate gale *or* near gale	32-38	50-61	whole trees sway; walking against wind is difficult
8	fresh gale *or* gale	39-46	62-74	twigs break off trees; moving cars veer
9	strong gale	47-54	75-88	slight structural damage occurs; shingles may blow away
10	whole gale *or* storm	55-63	89-102	trees uprooted; considerable structural damage occurs
11	storm *or* violent storm	64-72	103-117	widespread damage occurs
12	hurricane*	>72	>117	widespread damage occurs

*The U.S. uses 74 statute mph as the speed criterion for a hurricane.

beau geste \bō-'zhest\ *n, pl* **beaux gestes** *or* **beau gestes** \bō-'zhest\ [F, lit., beautiful gesture] (1914) **1 :** a graceful or magnanimous gest ture **2 :** an ingratiating conciliatory gesture
beau ide·al \,bō-ī-'dē(-ə)l, ,bō-,ē-dā-'äl\ *n, pl* **beau ideals** [F *beau idéa* ideal beauty] (1809) : the perfect type or model
Beau·jo·lais \,bō-zhō-'lā, -zhə-\ *n, pl* **Beaujolais** [F, fr. *Beaujolais* region of eastern France] (1863) : a light fruity red burgundy wine made from the Gamay grape
beau monde \bō-'mänd, -mōⁿd\ *n, pl* **beau mondes** \-'män(d)z\ o

In 1993 the new *Merriam-Webster's Collegiate Dictionary,* Tenth Edition, came out—and with changes from its version in the previous edition, the Beaufort Scale demonstrated that it's still evolving.

wires, and the umbrella that was "used with difficulty" is now "difficult to control." Most important, force 8 has gone from "breaks twigs off trees; generally impedes progress" to "twigs break off trees; moving cars veer."

This is important. On one hand, future dictionary trollers will find in the inclusion of cars the same thing I found in my little dictionary with the picture of the goat: "Sometime between 1982 and 1993," they'll say, "it became preposterous in the United States to try to describe any force of nature without including the effect it had on automobiles." That's kind of exciting.

On the other hand, this is just not the scale I fell in love with—not by a long shot.

Now remember, this is the Beaufort Scale of wind force, which as I eventually learned has been undergoing a long and genteel evolution since at least the sixteenth century, only eventually taking on Beaufort's name as a kind of protective coloration. I don't know why I should think that now it would have stopped changing. When I first opened the tenth edition, though, I had much to learn, and I presumed the Beaufort Scale was like the Celsius temperature scale or that metal bar a meter long that they have in a vault in Paris somewhere: a completed and unchanging metric against which we could hold up the world and compare.

The funny thing is, according to the World Meteorological Organization, who's in charge of it now, the Beaufort Scale *is* unchanging. It was most recently codified in 1970, and a helpful meteorologist at NOAA checked the 1974, 1988, and 1995 editions of the WMO Manual on Codes (that's WMO No. 306, if you're looking it up yourself) and found that the Beaufort Scale, including the shore criterion, remains unchanged.

Just not according to Merriam-Webster.

So I wanted to know how the scale made its way into the dictionary in the first place, and how the changes in the tenth edition had been made. An extremely languid correspondence with various people at Merriam-Webster eventually led me to Michael Roundy, the physical sciences editor, in charge, by the time I found him, of checking into things like the Beaufort Scale for the *Collegiate*'s eleventh edition, printed in 2003.

The reason for the table's inclusion into the dictionary in the first place, he told me, is lost to history, though his scant documentation did mention the terms appearing in the newspaper. The table at one point went up to 17 (including an extension developed in the 1950s) and also once included symbols for winds of different strengths, which were appearing on weather maps of the time. More important, the recent updates, he said, were prepared by a colleague no longer with the dictionary. Roundy guessed most of the changes were attempts to save words ("why specify an 'ordinary' vane . . . ?"). As for the inclusion of cars and the removal of telegraph wires, he regretted that his colleague had left no explanations for his changes. Roundy did agree, though, that "small-leaved trees" was a change in meaning, probably a typographical error resulting from the attempt to save words by saying "small, leaved trees." It was guesswork, though.

The trail didn't quite die, however. Mr. Roundy checked his own references and found a 1992 version of Gale's *Weather Almanac* that had a Beaufort Scale with some of the changes that ended up in the dictionary, including those moving cars; what Gale's source was, he didn't know. In a subsequent phone call to Gale's, I learned that neither did they—in fact, the current editor of the *Weather Almanac* had discovered the original 1906 scale online and had considered it an *update,* believing that because "generally impedes progress" did not limit observation to automobiles, it was an improvement over "moving cars veer"; I couldn't have agreed more. At that point, though, like Roundy, the Gale's editor shrugged and wished me well; he had no further suggestions. Roundy, though, did find it remarkable that the scale is not standardized, or that if it is, the standard is not adhered to.

Oh, and one more thing: That typo substituting "small-leaved trees" for "small trees in leaf"? He planned to make a change in the next printing of the eleventh *Collegiate*, though probably nothing more complex than replacing the offending hyphen with a comma. He thanked me for pointing it out. If there's a gold medal for old copy editors, they probably give it for finding errors in the dictionary. More than that, I had helped improve the Beaufort Scale in the place where it's most widely seen. As for the origin—or the fate—of "moving cars veer"? For now, none can say.

SO, AS THE SCALE SPREADS, IT IS HARDLY UNCHANGING—and, in fact, there's nobody completely in charge of it. And that in itself is kind of lovely. It enables changes like those I found in the tenth *Collegiate* to creep in, for the Beaufort Scale to be one of the ways we check not only the force of the wind but the force of change—when telegraph wires became overhead wires, when cars became ubiquitous. The Beaufort Scale itself becomes a kind of progress scale, a living history, recording not just the force of the wind but the details we observe to judge it.

It's no surprise something Beaufort propagated does this—in some way everything Beaufort touched is like this. While Hydrographer to the Admiralty Beaufort published "Abbreviations Used in Admiralty Charts," a guide to the symbols used in chartmaking—Beaufort, of course, standardized them. The first chart, from 1835, is a card about two-thirds the size of a sheet of copy paper, with a list of abbreviations: *c* means coarse, *ch* is chalk, and *crl* is coral. There are symbols for rocks just below sea level and rocks above sea level, specific types of lines that connect soundings of specific depths, a symbol for kelp.

By 1899 that sheet was eighteen inches square, with new symbols for floating lights, bell buoys, and "W.T." for a wireless telegraph station. By 1910 separate symbols could represent a light as among other things flashing, occulting, revolving, or alternating, and the chartmakers must have been enjoying the scenery: There were new symbols for glaciers and for lava. In 1947 radar stations, represented as little targets, appear, and by 1955 there are standard symbols for tunnels beneath rivers, submarine cables, and television towers. Nowadays the "Abbreviations" is Admiralty publication 5011, an eighty-page booklet in color, with an index. *Oz* still stands for ooze, and you still make the same symbol for kelp, but in its history the "Abbreviations" can stand as a marker for the introduction of dozens of technologies.

The Admiralty Charts themselves are the same. The Admiralty Chart for Plymouth from 1861, for example, includes a sketch of the harbor much like Beaufort's of Montevideo—it shows buildings, warehouses, the water filled with ships. That's been redrawn in 1881, with more ships—oddly, though even the 1861 chart included a steamship, the 1881 chart shows an increase in sail, not steam. In 1899 the new chart shows the astonishing sudden appearance on shore of Smeaton's Eddystone Lighthouse, reassembled overlooking the harbor when it was replaced in 1882. The harbor was resketched for the final time for the chart of 1954, showing a stark, socialist realism–style harbor of rounded smokestacks and neat corners—with no ships and no people. Soon more than just the people and ships had disappeared: the 1986 reissue of Chart 30, Plymouth, has no sketch at all.

Again, it's a beautiful timeline, providing all kinds of information—including exactly when the Hydrographer to the Admiralty figured, "You don't need a picture of the harbor, you've got machines to see it for you." It's a historical document, a progression. In a way, then, charts and their abbreviations become a record of their times, not just the harbor. The sketches and symbols on the charts record the details of their place and time in exactly the same way that *Ulysses* records the details of Dublin in 1904.

Like the Beaufort Scale they tell us about their users as well as their world. They're a monument to the art of description.

CHAPTER 7

Taking the Measure of the Wind: The Fabulous Beaufortometer

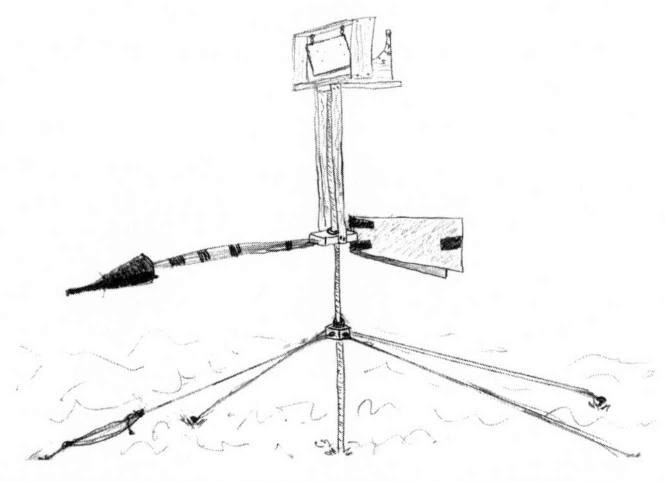

Unlovely but functional: the homemade fabulous Beaufortometer faced the wind and measured it.

STARTING A WIND TUNNEL IS a lot like starting a clothes dryer. You press a button and you hear a kind of grunt, then a metallic clang as high-voltage current magnetizes rotors, gears engage at high speed, and then there's a *whoosh* as air starts moving.

But that's the generator, which when it gets going sounds more like a subway train than a jet engine. In fact, the generators for the University of Michigan School of Engineering wind tunnel used to run the old Ann Arbor trolley lines. The entire system.

From the inside of the tunnel itself, should you find yourself standing there in gloves, sweatshirt, and goggles, you hear a very gentle *whooshing,* which becomes a sweep and finally a roar as the wind buffets in your ears. At about 30 miles per hour you can easily lean into the wind and remain standing; by 45 mph, you have to. While I was inside, the speed went up to about 65. It was no Hurricane Fran, but it corrected my notion that Daniel Defoe's description of a woman unable to crawl into the winds of the 1703 storm on her hands and knees was ridiculous. I ended up on my knees; I could crawl into the wind, but it wasn't easy. And I was wearing goggles.

I ended up in a wind tunnel because Perry Samson, my meteorological mentor, had obtained access for a few of his students. We brought along a home-built anemometer, which by then I was calling the fabulous Beaufortometer—a rickety-looking construction of one-by-twos, PVC pipe, plywood, pressboard, and a funnel. While the technician and I set up the Beaufortometer in the tunnel, Samson and his students made bets on how fast the wind would go before it blew my wooden contraption to pieces. They told me about the bets when I emerged (I could not stay in the tunnel during the test of the Beaufortometer; if it disintegrated, the fragments could be dangerous), and I said if it survived to its maximum—force 8—they would have to cook me dinner. Then we turned on the wind.

Samson and his wife eventually prepared crab cakes and a delicious soup.

I HAD BUILT THE FABULOUS BEAUFORTOMETER because, as I chased details about the history and trail of the Beaufort Scale, I feared I was losing track of the wind itself, the subject of those beautiful words I had come to love. So the first thing I did, to become familiar with the wind, was take a page from the journals of Sir Francis Beaufort, who kept a weather diary every single day of his life, and

A tiny handheld anemometer can let you know the exact, constantly changing, speed of the wind on a second-to-second basis. But then what?

keep my own wind diary. Once a day, I vowed, I would take a careful reading of the wind. I would check my surroundings, note what I saw happening, compare it with a tiny sixty-dollar handheld anemometer I had bought on the Internet, and write it down. The anemometer had a little propeller turned by the wind, and it measured wind velocity according to several units of measure—including the Beaufort Scale.

Unfortunately, once the little thing arrived I was instantly much more interested in measuring, to the tenth of a mile per hour (or the tenth of a knot, kilometer per hour, or meter per second), things like the air conditioning coming out of the vents in my house (about 10 mph) or the air rushing by my car window (pretty much the speed of the car; I suppose if I were driving with the wind it would be slower, but driving with my hand out the window was foolish enough, and I never stopped to do subtraction problems). I carried the little anemometer with me wherever I went, and almost the second it hit my hand, any thought of becoming a pure observer of my surroundings flew from my head. I jotted down things I saw, but I was careful always to connect what I saw with wind speed according to my tiny tool. I can thus tell you that by the time my eyes were open and I could check the gauge, a sneeze had turned the little rotor to around 19 mph, and that the air from the hand dryer in the men's room of the National Maritime Museum, in Greenwich, England, rages out of its screened fan at 71.5 mph, which is Beaufort force 11 and a mile and a half short of hurricane force.

This is probably not what Sir Francis Beaufort had in mind.

Or, then again, maybe it is. I was measuring my surroundings, using the tools at hand, and above all observing. I made little observations in my pocket notebook or in an old book titled "Memorandum" that June found in an old office she'd once inhabited. My resolution to record at least one wind observation every day lasted less than two weeks (just as well; to compare with Beaufort, I'd have had to keep going for seventy years), but I never did completely stop, and when I look at them now I appreciate them.

"July 6, Smithville TN, corner of W. Main & Public Square. Fiddle Fest. Glass wind chimes kept in regular motion by fan, but leap into bustle of action on wind gusts. Flag on corner flapping wide; flag in front of courthouse only lightly lifting. Anemometer 1.5–6.3 mph."

"Feb 4. Gas station, I-96. Snow flies horizontally, hard to push open car door into wind. Hard to pull open door to minimart. TV promised gusts of 45. These are not gusts. Flags fully horizontal, though not angrily. Car regularly buffeted by wind; snow swirls on pavement in irregular patterns. [Anemometer at home.]"

"My front stoop, July 8. Leaves on camellia vibrate, leggy flowers wave. Long blades of grass sway, but shorter ones stock still. 0–5.6 mph."

"April 1. Wind blows just strongly enough to make miniblinds buzz at a certain speed. Papers rustle on desk."

"June 6, Cumberland River park, Nashville. Wind extends

colored pennants. One-tenth-full Dasani bottle vibrates but does not tumble. Bubbles leave bubble machine sideways."

"September 10. Japanese lanterns on porch sway and swing—trees rustling in wind can be heard over traffic. Sustained gusts drown out neighbor's acoustic guitar. Riffles pages in magazines, flutters sheet music off lower shelf of end table near porch door. Wind felt on bare ankles."

"June 23. Ear rumble never high enough to drown out traffic noise, hair on legs rustles but not hair on arms. Shirt sort of stuck to back stays that way. 6.1 mph avg."

"August 4. Still. A million fireflies hang utterly motionless in the evening shade between two trees. Streetlight reflects perfectly in puddle—no jaggies."

These constant observations were getting me closer to what I had been chasing about the Beaufort Scale—it was about paying attention, and I was doing just that, if not particularly regularly. But part of what I had set out to do was to understand the wind as Beaufort had, and something wasn't quite there. So I determined that my measuring instrument—that little handheld anemometer—was not helping me. I wasn't watching the wind do something; I was reading a little LCD screen on which a number wavered as the wind varied. So I decided I had to build an anemometer that would enable me to see what the wind was doing simply by looking at it.

THE FIRST ANEMOMETER was built around 1450 by one Leon Battista Alberti, an Italian mathematician and architect. It was a pretty simple tool—a swinging plate that was deflected by the wind, much the way a giant shoe hanging in front of a cobbler's shop swings in the wind; people often speculate that Alberti based his anemometer on such a sign. His anemometer was something like a weather-

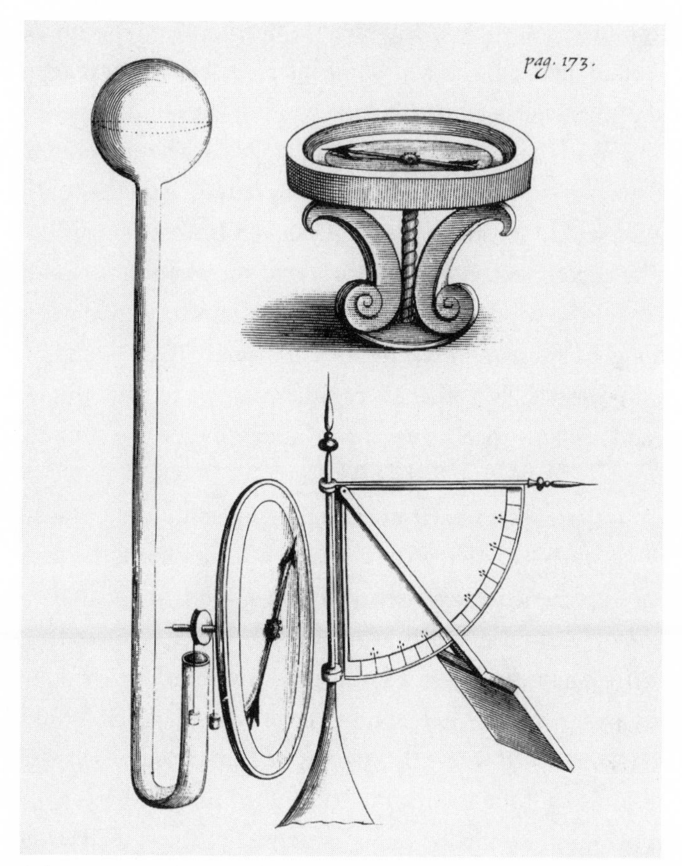

The deflection-plate anemometer (in the lower right corner of this drawing of early meteorological instruments) used by Robert Hooke around 1665 shows very little change from the swinging plate first invented around 1450 by Alberti.
From Sprat, *History of the Royal Society*. Special Collections Library, University of Michigan.

vane, with the swinging plate attached to the back, a scale along the arc behind it. In 1663 Robert Hooke of the Royal Society designed a simple version of Alberti's instrument with the scale on the inside of the arc, but little changed in anemometers until the 1800s.

Not for lack of trying, though. In the eighteenth century, what with the Fahrenheit and Celsius temperature scales, with growing understanding of the barometer (it hadn't been invented until 1643, almost two centuries after the anemometer), measurement was on people's minds, and they came up with some interesting attempts at getting the speed of the wind—at catching the wind by the tail, as it were.

William Derham, the guy who wrote to Defoe talking about degrees of storm, made an attempt to measure by letting "light downy feathers fly in the wind," according to one Alexander Brice, who tried to replicate the experiment in 1766. Brice found the method useless—"The velocity of the wind near the earth is very unequal, upon account of the frequent interruptions it meets with from hills, trees, and houses," he said, so he'd let fly feather after feather, and they'd whirl this way and that, and in all his trials he never saw "above one, or two at most, upon which I could have founded a calculation."

But he didn't give up. He noticed that higher up, above the houses, the wind "seems to be regular and steady." The clouds, he said, were "like a ship carried away insensibly by a smooth and gentle current, passing over equal spaces in equal times." So Brice marked a north-south line on his fields. On a sunny day he watched the shadow of the clouds pass that line, whereupon he listened to his clock count fifteen seconds, and noted where the shadow had advanced to. He measured the distance, did the math, and determined the wind was going 62.9 miles per hour. Brice tried this a few more times and then wrote it up for the *Philosophical Transactions.* His method isn't particularly practical, of course, but he did recognize an important fact—that the wind on the ground behaves much differently from the wind some feet up, which has a clearer path. Dropping feathers and watching clouds weren't the only unusual attempts at the developing science of anemometry, by the way. One weather historian described to me a British attempt to measure wind speed somewhat similarly by men with flags on poles. The men would run with the wind until the flag hung limp, which meant they were moving as fast as the wind was. They would then measure their own speed and determine that of the wind (the historian called it "athletic meteorology"). A meteorology teacher I know says he's done much the same when teaching kids, only using soap bubbles and having the kids ride bikes to keep up with the bubbles as they blow along.

A pressure-style anemometer was developed in France in 1721, and an improved one emerged in England in 1775, designed by James Lind, the physician who had also discovered that lime juice would prevent scurvy. This anemometer faced into the wind, which would exert pressure on a column of water in a U-shaped tube, causing one side of the water to descend and the other to rise, as measured along a scale. It wasn't particularly accurate—for one thing, until the wind was very high, the water was moving tiny fractions of an inch. In addition, like the swinging plate, it wouldn't be much good on a pitching and rolling ship, which was of course one of the reasons something like the Beaufort Scale was needed. (In *Practical Navigation,* Dalrymple himself said, "It would be a very great Improvement in this part of Nautical Knowledge, if the Instrument, for registering the Wind's Velocity, was introduced into common use at Sea.") The spinning-cup anemometer with which we are familiar today was invented in Ireland in 1846 by Thomas Robinson, and William Dines improved the pressure-tube anemometer around 1900, inventing one that measured both the force and the direction of wind.

In a way, though, the Dines and Robinson anemometers were already distancing the observer from the wind itself. The pressure tube required the observer to pay attention, instead of to the world around him, to tiny movements in a column of water; the spinning-cup anemometer required a mechanical or electrical counting mechanism, since by force 4 or 5 the cups spin too fast to count their revolutions by eye.

IT'S ONLY GOTTEN MORE COMPLEX. I spoke to Frank Marsik, a University of Michigan professor who specializes in meteorological instrumentation, and in a room filled with ancient instruments like pyranometers (the glass globes that measure solar radiation) he pulled out a sonic anemometer that costs around $20,000 and looks, with six little nodes suspended in space by metal rods, like some kind of deformed octopus. Ten times per second it measures the temperature and the speed and direction of the wind. This isn't a bad thing—for high-level science it's necessary. But Marsik points out that the more accurate the anemometer, the more it demonstrates the wind's fundamental unmeasurability. It's never blowing 20 miles an hour; it's 20.4, then 19.1, then 17.5, then 22.6. "It's amazing," he says. "Just how dynamic, how constantly changing the atmosphere is." The wind is gusty and changeable by its nature, so in a way anemometers themselves are all wrong for the job. The literature of wind measurement is full of people writing about different ways to measure wind speed through effects: "Chicken Plucking as a Measure of Tornado Wind Speed"; "Wind Speed Estimation Based on the Penetration of Straws and Splinters into Wood"; "The Everyday Effects of Wind Drag on People." What's needed is something general—something like the Beaufort Scale.

Once I had followed anemometers through their development, I knew what I wanted to build. I wanted an anemometer like the one Alberti would have had. I wanted to measure the wind like the first people who measured it—I wanted my anemometer to be as physical as one of those hanging signs that were its supposed inspiration.

You can find all kinds of makeshift anemometer projects on the Internet and in books, commonly involving straws and Dixie cups. But in a book from 1947 called *Techniques of Observing the Weather* (by B. C. Haynes, chief of the U.S. Weather Bureau observations), I found general plans for a plain old swinging-plate anemometer. They were easy to adapt: A wooden frame, from which hung a 7-ounce plate of 72 square inches. A piece of plywood with an arc cut out supplied the scale, which measured the wind in Beaufort numbers—force 2 was at

4 degrees, force 3 at 15, and so on, until at force 8 the wind would push the plate to 81 degrees, nearly horizontal. I mounted the assembly at the top of a PVC pipe on a bearing I bought from an auto-parts store, so it would rotate freely; out the back I mounted a split-wing vane (the book recommended the wings spread 22 degrees, and that was good enough for me); out the front, to balance it, I put a dowel, with a funnel mounted on the front, for an arrow. Getting it to stand was a project—but I took a cue from the sailing ships of Sir Francis and instead of fighting the wind I worked with it, using ropes to rig the thing, rather than wood to support it. With four guy ropes, whichever way the wind was blowing, at least one rope was steadfastly holding it up. Solving the building problems was half the fun, and when it was finished, at about seven feet tall, it looked like a cross between a mailbox, a poorly conceived clothesline, and the Tin Man, but once set standing I thought it had a certain unselfconscious elegance. (I could also, Haynes's book told me, have made a barometer—out of a sealed coffee can—and a hygrometer, measuring atmospheric moisture content, out of human hair. Blond works best.)

More important, the Beaufortometer worked. It not only stood the test of the wind tunnel, it withstood the much greater assault of actual wind and rain. I stood it up in the backyard and planned to enjoy watching it until it crumbled. I did enjoy watching it—when the wind would kick up it would, in a stately manner, seem to wake up, seek the wind, and finally point into it, and the plate would swing out, gently but perceptibly. The wind tunnel had shown that it was surprisingly accurately calibrated, so if I wanted to, I could have kept a wind diary using it as my metric.

But I never did.

I FOUND, SURPRISINGLY, THAT ACTUAL WIND SPEEDS meant nothing to me. That is, I went in exactly the opposite direction from the scientists. Once the Beaufort Scale—or wind scales like it—became widespread, the scientific and meteorological communities wasted no time in trying to pin the wind down numerically. Force 8 might mean something to sailors, but meteorologists trying to use the vast amounts of data collected around the world to begin understanding how the atmosphere worked, needed actual wind speeds, and as accurate and dependable anemometers appeared, researchers made trial after trial to assign specific velocities to the Beaufort Scale. The scale, it seemed, had begun to outlive its usefulness—as anemometers could measure the wind to within a tenth or hundredth of a mile per hour, something as general as the Beaufort Scale would lose its scientific value. In any case, committees continued to try to pin down the wind speeds of the Beaufort Scale, and trials like these created the articles in 1874, 1896, and many others comparing and organizing wind scales. And, of course, the greatest of these, organized by the Met Office, finally yielded the 1906 Publication 180, in which the descriptive Beaufort Scale as we know it now was written by Sir George Simpson—and a committee.

That description still held my interest. I noticed that whether I was using my tiny handheld anemometer or my seven-foot Beaufortometer, I was less interested in the actual speed of the wind than in the way the wind affected the world, and the way I wrote down what I had seen. My wind diary was much more interesting to the degree that paying attention to the wind focused my attention on the world rather than on some meter. Fine, the July wind in Smithville blew up to 6.3 mph; but much more important, that was strong enough to make the wind chimes riot and make the corner flag flap. It didn't tell me much on a still summer night that the anemometer didn't even spin; but the way the pendulous abdomens of fireflies hung down in the motionless air beneath the Nashville pines? The way their lights reflected off the perfectly unruffled puddle surfaces? That felt like seeing something, like truly perceiving my surroundings. I was the meter, the Beaufort Scale seemed to imply; that, somehow, was the point.

And not just for me. As increasingly accurate and increasingly sensitive meters took hold, in science and in meteorology, descriptive scales weren't, after all, disappearing. In 1870, for example, for those observers at the ends of its telegraphic network without anemometers, the Smithsonian Institution published *Directions for Meteorological Observations.* It returned to the five-level (0–4) scale used by Jurin and Celsius and the Palatine Society, with the addition, between 0 and 1, of a single letter for the direction of the wind, which "means a slight movement of the air hardly to be called a wind, and only just sufficient to allow an estimate of its direction." It goes on through 4, but it gives extremely thorough descriptions: 1 is "a light breeze which moves the foliage, and sometimes fans the face"; 2 moves branches of trees and "somewhat retards walking, and causes a slight rustling sound in the open air"; 3 already causes "entire trees to rock, . . . and carries light bodies up into the air"; by 4, "trees are in constant motion; . . . now and then chimneys, fences & c. are thrown down." The Smithsonian estimates that 1, 2, 3, and 4 of its scale are equivalent to 1, 4, 8, and 11 of Beaufort's scale, showing that they were, if not willing to adopt the international standard, at least aware of it. (To see most of the variations on the Beaufort Scale discussed here all in one place, please refer to Appendix A.)

And, clearly, the Met Office was aware of the Smithsonian's work. "Trees . . . in constant motion" is a phrase that moved almost unchanged into the famous Met Office 1906 scale, making even clearer that that Met Office version of the scale that meant so much to me had an even broader background than I had expected. From Tycho's first scale, with all the crazy names of the winds, through all the sailing scales, from Jurin's five-level scale to the Celsius scale to the million other wind scales that propagated out of control in the nineteenth century, the Met Office had plenty of influence as Sir George Simpson gathered his committee and his observations and created the 1906 Beaufort Scale, with its poetic language and vigorous description.

AND THAT SCALE WAS FAR FROM the last of such scales, as it happens. Perhaps the greatest other descriptive scale had been designed only a few years earlier, itself part of a long evolution. In 1874 Michele Stefano Conte de Rossi, an Italian geologist, published a ten-level descriptive scale of earthquake intensity; independently, in 1881, a Swiss named Françoise-Alphonse Forel designed a similar system. Their combined scale—the Rossi-Forel Scale—began to gain international currency in the late 1800s. In 1902 it was improved by the Italian geologist Giuseppe Mercalli, whose name has inexplicably remained attached to it ever since (probably because, like Beaufort, he spread it around effectively). Mercalli's scale was expanded to twelve categories by a guy named Cancani, who used only single words ("weak," "damaging," and so forth) on his scale. In 1912, though, the German August Sieberg filled out the scale, adding complete descriptions. Called the Modified Mercalli Scale of Earthquake Intensity, it was translated into English in 1931, and it resembles nothing so much as the Beaufort Scale.

No surprise—it's designed to solve exactly the same problem the Beaufort Scale did: to help people, in the absence of useful or trustworthy meters, to quantify a natural force by relying on their perceptions. The *Manual of Scientific Enquiry,* remember, suggested grandfather clocks as earthquake sensors; by the third edition, it also suggested a series of cylinders of different diameters, which would tell the strength of the quake by which ones fell over.

Mercalli and his cohorts chose Beaufort's solution: He simply described what would happen in different situations. Just like the Beaufort Scale, there are now at least half a dozen versions of the Mercalli scale. In the 1931 translation, a level I earthquake is not felt except "under rare circumstances," though "sometimes birds, animals, reported uneasy or disturbed; sometimes dizziness or nausea experienced; sometimes trees, structures, liquids, bodies of water, may sway—doors may swing, very slowly." By level III, people would report "vibration like that due to the passing of light, or lightly loaded trucks, or heavy trucks some distance away." A level V earthquake had a lot more oomph: It "awakened many, or most. Frightened few—slight excitement, a few ran outdoors. Buildings trembled throughout. Broke dishes, glassware, to some extent. Cracked windows—in some cases, but not generally. Overturned small or unstable objects, in many instances, with occasional fall. Hanging objects, doors, swing generally or considerably. Knocked pictures against walls, or swung them out of place. Opened or closed doors, shutters, abruptly. Pendulum clocks stopped, started, or ran fast, or slow. . . . Spilled liquids in small amounts from well-filled open containers."

This is by no means as poetic or concise as the 1906 version of the Beaufort Scale—that full description of level V, for example, is 108 words on its own, two words fewer than the entire 1906 Beaufort Scale. Just the same, it's great descriptive prose (level VI: "small bells ring"; level VIII: "alarm approaches panic. Disturbed persons driving motor cars. . . . Twisting, fall, of chimneys, columns, monuments") and does its job as effectively, if not as elegantly, as the Beaufort Scale. And it too changes through time (it's been updated through the years) and location (a Turkish version describes minarets, not columns, swaying and falling). There are other earthquake intensity scales, most of them descriptive ("rattling of doors and shoji [Japanese sliding doors]," says the Japanese Meteorological Agency scale of force 3). Then of course there's the famous Richter Scale, designed in 1935 by Charles Richter.

It's logarithmic, which means an earthquake of 6 is 10 times as strong as a 5, which few average people understand, and in practice means that news media describe only earthquakes from about 3.5 (about the size of the smallest earthquake to which anyone pays much attention) to 7 (a quake that causes serious damage). The Richter Scale, which has no upper or lower limit, is highly useful to scientists but conveys little actual meaning to most people. A switch to description would convey more information more effectively.

These descriptive standards all do exactly the same job as the Beaufort Scale when it was finally spread by Sir Francis: it took something well worth measuring, but for which there were not yet accurate—or useful—metrics, and used instead of a mechanical device that greatest of all meters, human perception.

THE FINAL DESCRIPTIVE SCALE TO which my pursuit of Beaufort led me was a new version of the Beaufort Scale itself. Not an update of Beaufort's list of sails used, not an improvement of Simpson's 1906 shore criterion, but a completely new gauge, and it's been so successful that many reference books cite it as the Beaufort Scale with no further explanation.

As the 1906 scale recognized, steam power for ships was rendering Beaufort's amount-of-sail method useless. Beaufort had worked hard to find something on which sailors could base their estimates of the wind, and had settled on the structure of the ship itself, but as steam power improved, ships carried fewer sails and often, finally, none at all. In Germany, from 1903 through 1905 (the same period, coincidentally, during which Simpson's committee did its work), two German captains named Prager and Petersen made simultaneous Beaufort Scale estimates and observations of the state of the sea, finally publishing in 1927 what is now called the state-of-sea scale or, sometimes, Petersen's scale. It includes descriptions for every number on the Beaufort Scale, starting with of course 0, Calm, which is "sea like a mirror."

In 1, Light Air, "Ripples with the appearance of scales are formed"; by 3, "crests begin to break. Foam of glassy appearance. Perhaps scattered white horses." Things move along to 6: "Large waves begin to form; the white foam crests are more extensive everywhere. Probably some spray." In 8 "edges of crests begin to break into spindrift," in 9 "crests of waves begin to topple, tumble, and roll over," and in 10, Storm, "the 'tumbling' of the sea becomes heavy and shock-like." In 11 the waves are exceptionally high, and, terrifyingly, "small and medium-size ships might be for a time lost to view behind the waves," which made me rather glad my time aboard the *Europa* had seen so little wind. In 12, Hurricane, "The air is filled with foam and spray. Sea completely white with driving spray; visibility very seriously affected," to say nothing of one's courage or will to live.

Again, the state-of-sea scale is hardly a match for the 1906 shore criterion in conciseness, but it's lovely and thorough, and it surely gets the point across. You wouldn't call it an update of the Beaufort Scale—it's more of an adjunct, a translation. It's the filigree at the end of the Beaufort Scale, chocolate frosting on chocolate cake—something just as satisfying and basically the same that adds an extra touch of sweetness, of delight, to the whole undertaking. In addition, it's probably the most widely dispersed version of the Beaufort Scale in use today—in the U.S. National Weather Service Observing Handbook No. 1, used aboard ships participating in any of the service's weather programs, the version of the Beaufort Scale included is the Petersen state-of-sea scale.

The rest of the history of the Beaufort Scale is rather plain. Officially adopted by the International Meteorological Conference in 1874, it continued to assert itself as an international standard. In 1912 the International Commission for Weather Telegraphers got to work on an updated standard, though World War I interrupted its efforts, so it wasn't until 1926 that the updated wind speeds and descriptions from the 1906 scale were adopted as an international standard. The International Meteorological Conference in Warsaw in 1935 asked several countries to test the Petersen state-of-sea scale, and it was finally adopted in 1947 during the International Meteorological Conference in Washington. That conference also endorsed the change of the International Meteorological Organization, dormant during World War II, into the World Meteorological Organization, an office of the United Nations.

Among its first adopted standards was the Beaufort Scale, adopted in its final version in 1970.

NOT THAT THAT WAS AN END to the scale's development. As I discovered in the dictionary, despite being an international standard, the scale changes with time, almost organically. In 1974 Herbert Saffir, a Florida engineer, and Robert Simpson, of the National Hurricane Center in Miami, prepared a Hurricane Disaster Potential Scale of five levels, which has joined the Beaufort Scale now (you sometimes find it stretching to 17, with the five Saffir-Simpson levels added). Those five levels are, of course, the basis of the "Category 3 hurricane" that you hear from the Weather Channel all fall every year.

And they too are purely descriptive. Category 1 specifies "Damage primarily to shrubbery, trees, foliage, and unanchored mobile homes." By Category 3, "Practically all poorly constructed signs blown down. Some damage to roofing materials of buildings; . . . Mobile homes destroyed. . . . Low-lying escape routes inland cut by rising water three to five hours before hurricane center arrives." By Category 5, the storm surge does "major damage to lower floors of all structures less than 15 feet above sea level within 500 yards of shore. . . . Massive evacuation of residential areas," and you should expect "complete failure of roofs on many residences and industrial buildings; . . . Small buildings overturned or blown away. Complete destruction of mobile homes."

Poetry it's not, but straightforward description no doubt, with an obvious debt to Sir Francis. But hurricanes are hardly the only other wildly describable winds; if the Saffir-Simpson Scale sounds a little like the "mobile home guide to hurricane intensity," that brings up tornadoes, and—did you doubt it?—there are a couple of tornado intensity scales too.

The first was written in 1971 by the University of Chicago physicist Theodore Fujita and is called the Fujita scale (sometimes Fujita-Pearson, including Allan Pearson, a meteorologist who helped develop it). F0, a "gale tornado," has winds of less than 73 mph and causes "some damage to chimneys . . . shallow-rooted trees pushed over." By F2, a "significant tornado," winds are between 113 and 157 mph, "mobile homes demolished; boxcars pushed over; . . . light-object missiles generated." Work up through "severe" and "devastating" tornadoes to F5, "incredible tornado," and the wind climbs to 261–318 mph, in which "strong frame houses [are] lifted off foundations and carried considerable distances to disintegrate; automobile-sized missiles fly through the air 100 yards or more; trees debarked." Like the Mercalli scale, the Fujita-Pearson scale does its best work in pure description. As it happens the scale includes wind speeds, but the descriptions are what enable observers to estimate those speeds (which are simply too fast to measure, because winds at that speed destroy meters). And actually, according to the National Climatic Data Center, the Fujita scale functions as a bridge between the Beaufort Scale and the speed of sound—F1 starts, not coincidentally, with the highest wind speed on the Beaufort Scale (73 mph), and there are Fujita numbers up through F12, which is 738 mph, the speed of sound.

Naturally, once something that useful was developed, someone else developed a competing version the next year—the Torro scale, developed in England by Terence Meaden. It's pretty much the same as the Fujita scale except it has two levels below Beaufort 12 ("light": "marquees seriously disturbed; . . . trail visible through crops"; "mild': "deck chairs, small plants, heavy litter become airborne"). The Tornado and Storm Research Organization, an independent enthusiasts' group founded in 1974 by Meaden, has also come up with a Torro scale for hailstorm intensity ("H1: Leaves and flower petals punctured and torn"; "H4: Some house windows broken; . . . birds killed").

And everywhere you look, you can find other scales: the Mohs hardness scale (ten steps, from talc to diamond; you measure by seeing what is hard enough to scratch something else); the Scoville scale for pepper hotness; and the Torino scale that measures, from 1 to 10, how likely an extraterrestrial body is to crash into the earth, and how much damage it will do if it does. There are scales for neonatal development, for pain assessment, for determining whether you're in a coma or not. Medicine is full of scales like this, which makes perfect sense: Medicine is full of doctors and nurses trying to get a sense of the seriousness of a problem—such as pain—that can't be measured by machines: It can be communicated only in language. When I spoke with a nursing professor about the Beaufort Scale, she smiled and shrugged: "It's a diagnostic tool," she told me. "We use those all the time."

THESE OTHER SCALES ARE IN EVERY WAY like the 1906 version of the Beaufort Scale. They're brilliantly lucid description, the universe we can see, hear, and feel distilled to its essential oils.

Oh yes, and we can update them, too. Once I realized, courtesy of Merriam-Webster, that the Beaufort Scale changes as it moves forward through time—and that despite my love of the wording of that

1906 scale, that change is a perfectly reasonable thing—I began looking for updated, personalized Beaufort Scales. And they're not hard to find.

Alisa Crawford, the miller at *De Zwaan,* found a kite-friendly version of the Beaufort Scale in a kite catalog (much below force 4 or above force 6, it showed, wouldn't be much good for flying kites). One in *Yachting Magazine* was printed in 1966 probably more as a public service than anything else—the page looks designed to be taped on a bulkhead—but the publishers added a column of sailing conditions: "fine sailing but wet" for force 5, "all boats better off in harbor" for force 8, and if you didn't get the message, "this is no time for yachts to be at sea," for all numbers higher than that. One sailing website added a column for "approximate effect on racing dinghy": for force 2, "crew sit on windward side of boat"; by force 4, "crew ballasting out hard over weather gunwale." That scale even includes a "psychological scale," running from "boredom" at 0 to "great pleasure" at 4, reaching "anxiety tinged with fear" at force 7, and "panic" at force 10.

There's a scale designed for people who sail in canoes: "Don't bother to put sails up; paddle," it says of force 0, which it calls "not worth it." By "pleasant," force 2, "sailing downwind is faster than paddling but paddling upwind would probably be faster." Force 5 it calls "too exciting!" and describes as a "canoe-sailors gale. Almost survival time if caught out." And of course *Punch,* the iconic British humor magazine, periodically runs jokey Beaufort Scale revisions: "All visitors off course. Foreigners complain of British Summer," says a 1960 version, designed for seaside resorts, of Force 10; one from 1981 updates the scale for among others female cyclists ("Force 8. Gale. Hairdo completely ruined") and lawyers ("Force 7. Near gale. Wonderful series of industrial accidents and collapses").

I love these personalized, updated Beaufort Scales, and I think

Beaufort would approve: They're communication, ways to get the point across. But of all the improvised Beaufort Scales I've found, my favorite by far is the scale as it appeared in July 2002, on the website savannah-weather.com, based in Savannah, Georgia. Along with the Fujita-Pearson Tornado Intensity Scale and the Saffir-Simpson extension of the Beaufort Scale, savannah-weather.com provides a Beaufort Scale including simple, reduced versions of the Peterson state-of-sea scale and the 1906 shore criterion. But instead of merely including the scales, the people in Savannah added elements they thought would improve the scale for their readers.

They were completely right.

In Force 1, where the 1906 scale says "direction of wind shown by smoke drift but not by wind vane," the Savannah scale adds, "Spanish moss sways." By Force 6, where the 1906 scale says "telegraph wires whistle," Savannah adds, "wind heard in pines." By Force 7, that overheard wind is a roar. Best of all, after 7, this scale becomes a lawn furniture scale: Force 8: "Lawn furniture rocks"; Force 9: "Lawn furniture moved"; Force 11: "Lawn furniture airborne."

This—this is the essence of the thing. This is what Beaufort wanted: A *useful* scale, based on the objects with which one regularly comes in contact. This scale is every bit as useful to the people in Savannah as the sail scale was for sailors, as Smeaton's windmill scale was for millers. In Savannah, people are out in the yard, or in the park, all the time. Lawn furniture is by every beach house and on every porch, Spanish moss hangs from live oaks, tall pines catch the bottom of the winds aloft. Those are what people will notice; those are what will help people recognize the wind. This is like the *Abbreviations Used in Admiralty Charts,* pegging themselves in time by including telegraph offices, or radar installations: There are undersea cables but not yet radio towers—this must be the turn of the twentieth century. This version of the scale locates itself in space by noting its surroundings: lawn furniture and Spanish moss—it's the American South.

This is what the *Manual of Scientific Enquiry* would want—people observing what they see, noting it, sharing the information, trying to communicate with one another about it. The Beaufort Scale is alive, moving forward through time with the rest of us, but Beaufort's spirit inhabits it every time it moves. *Spirit*—the word comes from the Latin *spiritus,* breath, that most basic wind.

THE WIND. The more I looked into scales and their history, the more I sometimes forgot that the basis of all this discussion and clarification, this categorization and communication, was the wind, the invisible, ineffable wind. But I loved the Beaufortometer—I loved to watch it swing its arrow around in the yard, watch its plate rise and fall with the wind—and the more it drew me back out to the world, the happier I was. Even when I was failing—as I most often was—to continue making entries in my wind diary, I found that I was paying closer attention to the wind than I ever had before, and I never regretted it.

One way I worked on being aware was by getting in the habit of carrying a tiny pocket kite that I bought. A parafoil kite, it weighed an ounce or so and would fly in what I estimated to be winds of force 3 or above. If the trees were shaking, there was usually enough wind for my kite to at least fitfully rise. And one of the things I learned from the kite is that Alexander Brice was right—the air along the ground and the air higher up are moving in different ways, at different speeds. Sometimes it was clear from the clouds and the treetops that there was a nice breeze up there, but if I was in a hollow or behind buildings or a rise, I couldn't get to it, and the kite never got anywhere. Sometimes it was the opposite—a nice brisk breeze would be flowing along the ground, but the kite never wanted to go any higher than forty feet or so, above which the air seemed much stiller. It turns out that the friction of the ground plays a great role in the course of the wind along the ground—the wind is actually sort of falling forward as it moves along, catching itself on trees and hills, houses and buildings. Between buildings it can funnel in and create gusts that rush through the narrow space much faster than the wind itself—think of putting your finger on the end of a garden hose. I would often hear the hollow rattling of bamboo wind chimes on a neighbor's second-floor balcony, meanwhile standing in air perfectly still. The wind is like an ocean current, like a motion of water through other water—the atmosphere is often called "an ocean of air," and the winds are its currents, responding to obstacles just the way water does in a river. There are entire careers in meteorology—and architecture—dedicated to the behavior of the wind as it moves in the 30 meters or so right along the face of the Earth. Most meteorology services now try to measure the wind at 10 meters above the ground for that reason. Then you're measuring the movement of large parcels of air, not gusts caused by you and your neighbor building your houses too close together.

But down by the ground is where the fun is, and anyhow, that's where we live. Down there at the bottom of the wind I found myself looking for its currents. I would notice the way dandelion seeds, those little puffs of fiber with a tiny brown stalk attached, blew off a dandelion when the wind reached just a bit over force 1. I would squat in a vacant lot or on someone's lawn and see them depart as a group and then, almost miraculously, glide along tiny currents among stones and grasses as the wind made its way along the ground. They were like helicopter squadrons, like dragonflies, with their tufts and their tails, and the steadiness of their progress was stunning. Proud as we may be of our windmills, sailing ships, airplanes, the dandelions have adapted to the wind better than we have.

Trees, too. Cottonwoods release their seeds in the spring, and if you stand in the right place you can feel that you're in a snowstorm, as thousands of thumbnail-sized puffs release from trees in the slightest gust, whirling and drifting as the breeze progresses. When they land, they roll along asphalt like tumbleweeds, but when they encounter anything organic—dust, dirt, sand, another plant—they entangle, ready to take their shot at becoming a tree. Maple trees, too, line the streets where I live, and in seeking the haunts of the wind I saw that their seeds work differently than I had always thought. They don't spiral down from the tree, trying to catch a ride to fertile ground. Instead, they drop below the tree, all of them, over the course of a couple weeks in spring. Only after they've fallen do they dry out, lighten, and separate from one another. Then, when one of those gusts of wind comes, they leap from the ground, moving in every direction, and using their helicoptering spiral to land far away from one another. Their motion provided yet another way to watch the wind.

But I'm hardly a master at that. For biological wind watching, the master is a man named Lyall Watson, who, in a book called *Heaven's Breath,* created what he called "the first Biological Beaufort Scale." He goes only from 0 to 10, probably figuring like Robert Vos, captain of the *Europa,* that beyond that you're just hanging on anyway. He lists the numbers and names of the scale, then creates columns for human activity, whole plants, seed & leaf, birds, and invertebrates. At force 1, when the 1906 scale tells us that "direction of wind shown by smoke but not by wind vanes," Watson tells us in "seed & leaf" that "light plumed seeds airborne"; in "birds" that there are "thermals with many soarers"; and in "invertebrates" that "Aphids fly, spiders take off." When those maple seeds glide, he says, it's force 3 or higher, by which point "hoppers, aphids, spiders grounded." Your clothing flaps at force 4, he says, and force 6, strong breeze, is strong enough that "arms [are] blown out from sides." By that point "few small perching birds [are] in flight," and even moths and bees are grounded.

He calls force 7 the "biological wind threshold," noting that even Beaufort makes the change there from breeze to gale, and that weather bureaus begin issuing small craft warnings then. By force 9 "children [are] blown over," and at 10 adults are, and there he stops. Once the wind is knocking you over, how much further do you need to go?

IT'S BEAUTIFUL, IT'S USEFUL, IT'S EVERYTHING a Beaufort Scale should be. So I was hugely jealous of Watson's scale. True, I had caught a copyediting mistake in the Merriam-Webster version of the scale, and there's honor in that, but that had more to do with the dictionary than the Beaufort Scale. I wanted to add something to this scale, and I looked for a way to do so.

Writing scales, of course, is easy—as the psychologists say and as I found as I looked around, it's a fundamentally human thing to do. We organize, we rank, we categorize. We look at everything and compare it with everything else, lining things up for easy reference. In fact, I began thinking of things in scales—mail would come and I would rank it on a correspondence scale, a scale of epistolary satisfaction: 0 would be one of those notecard fliers about cable rates, something that a person didn't touch from the moment it was designed until the mail carrier dropped it off; 1 would be junk mail that at least came in an envelope, 2 or 3 a catalog from a company you didn't mind leafing through. An actual letter with an actual stamp instead of a sticker would get a 4 or 5; one with something interesting in it—a photograph? attractive stationery?—would go higher. An international stamp would bring a letter up to the high single digits, and I figured 11 would be a handwritten letter from your favorite person, with a real stamp, with something extra inside that you weren't expecting. Twelve, it would be all that plus perfume.

Or something like that. It was a fun way to think, to get used to organizing my perceptions. But it wasn't anything anybody needed, it wasn't solving a problem. I wanted to do as Watson did, to add something to the Beaufort Scale of wind force, to take this useful thing and find further use for it—to take my own descriptions and add them to the pile. I would stand on the porch and watch the trees and I would think about it, or I would take walks. Sometimes I would stop and have an occasional smoke while I did so. That's bad and cigarettes are poison, I know this, but still that's what I did. I did that often enough and thought about it long enough that eventually I did come up with a scale. It's not particularly useful—in fact I hope nobody ever uses it—but, when all was said and done, I felt like I had followed Beaufort and his ilk and used my observations to create an actual scale, and it was useful, and it had to do with the wind.

So here, from my notebook, is the world's first Beaufort Scale of Outdoor Cigarette Lightability.

I DON'T CLAIM THIS WILL PASS INTO GENERAL USE, and I certainly forbid anyone from picking up a cigarette to try it out. Still, it does what a scale sets out to do—it classifies observation according to results, and it presents those results in a format that makes them easy to consult and use. If I took a walk to have a smoke, and when I got home my girlfriend wanted to know what the wind was like, I could tell her.

I can't claim that it's poetry, but it's in the spirit.

0	Calm	You can light a single match, then gesticulate around like a moron while making some pointless assertion, then raise the match to light the cigarette and the match is still lit.	
1	Light Air	Or that in which a reasonable person, on a city street, could lean over and between cupped hands light a cigarette using . . .	1 to 2 matches
2	Light Breeze		3 to 4 matches
3	Gentle Breeze		5 to 6 matches, but might not be possible
4	Moderate Breeze	Match works only in doorway; otherwise lighter needed to get cigarette burning.	
5	Fresh Breeze	Butane lighter needed; Zippo still works in store alcove.	
6	Strong Breeze	Extra hands needed; butane lighter rattles in wind.	
7	Moderate Gale	Butane lighter rattles even in arcade, if no doors present.	
8	Fresh Gale	Step inside, Clyde.	

CHAPTER 8

A Picture of the Wind: Poetry, the Shipping Forecast, and the Search for the North Shields Observer

IF YOU STAY UP LATE ENOUGH in England and you listen to BBC Radio 4 at around ten to one in the morning, you have one of the greatest auditory experiences this planet has to offer.

You hear the Shipping Forecast.

It's actually broadcast four times a day (about 6:00 A.M. and about 1:00 A.M. on Radio 4–FM, about noon and 6:00 P.M. on AM), but the one people rave about is the 1:00 A.M. broadcast. The format is the same each time: Music—a mincing string-and-flute instrumental called "Sailing By"—plays for a good minute or more. Then you hear the voice of a newsreader, who gives a brief general synopsis of the weather and then proceeds clockwise around the British Isles, listing the weather forecast for each of thirty different defined sea regions surrounding England. Some have homey names, like Thames, Dover, or Wight; others sound unfamiliar or remote: Viking, German Bight, Southeast Iceland. In the patented modulated tones of BBC News, the presenter lists the name of an area (or more than one, if the forecast

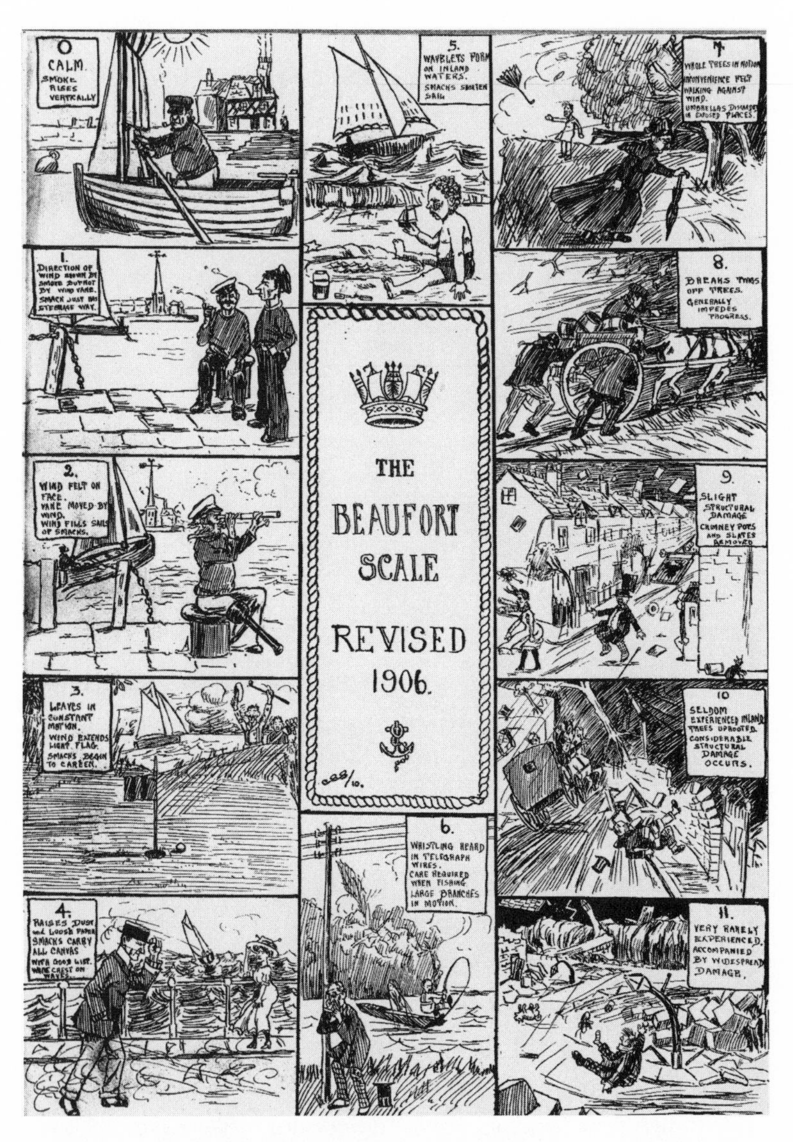

The name of the author of this cartoon has been lost, but the drawings show how powerfully the language evokes images. The cartoon was drawn in 1906, the year the shore criterion first appeared in Met Office Official Publication No. 180.

Met Office, National Meteorological Archive. Used by permission.

for adjacent areas is the same); the wind conditions; the general weather expected; and the visibility: "Tyne, Dogger: West, 3 or 4, becoming cyclonic, then north 4 or 5; occasional rain; moderate or good." Then the next area, in clockwise order. The whole forecast takes about ten minutes and is one of the most beloved radio traditions in England. And of course those numbers, gloriously undefined as they cascade out of the radio in what amounts to a chant, a Zen poem, a mantra, are nothing other than Beaufort Scale numbers.

Because of the Shipping Forecast, most people in the United Kingdom know the name of Francis Beaufort. The Shipping Forecast was first broadcast in 1924 and, with the exception of its suspension during World War II, has been a BBC staple ever since. "Sailing By" was written in the early 1960s, originally for a television feature about ballooning, but it's now inseparable from the Shipping Forecast. (You can hear the Shipping Forecast online at www.bbc.co.uk/weather/ukweather/shipping.shtml.)

When I first started asking around about the Beaufort Scale, every time I mentioned it to someone from the UK, they would nod, and a little smile would creep across their face. "Ah, like in the Shipping Forecast," they'd say, and follow with a description of themselves, lying snug in bed, the radio playing; the book is closed, the lamp off—then those first prissy strains of "Sailing By," then the even, reassuring tones of the presenter, and then, for ten peaceful minutes, a stream of words and numbers. If you remain awake for the entire ten-minute recitation (the noon and 6:00 P.M. versions are only three minutes, by the way), at 1:00 A.M. you hear "God Save the Queen," which renders the whole thing even more exquisite, but according to most of the people I spoke with, they rarely hear the end. They lie in bed thinking of the gale in Dogger or the wind veering in Shannon, they appreciate their own safe surroundings, and to the reassuring intonation of numbers and nouns, they fall asleep.

"It's like a lullaby almost," said Charlotte Green, who for many years was the presenter who read the late-night Shipping Forecast. "It goes far beyond names and numbers. It's quintessentially British. It couldn't be that way anywhere else. The sea exerts a huge influence on the British psyche and imagination." The forecast brings that up, Green said, "the waves and the huge unbounded size of the ocean." She said she did think of her listeners in bed, thankful for their comfort, but she especially thought of the people for whom the forecast was intended, particularly when the forecast was rough: "You really would feel for the sailors who were out there struggling to take down what you were saying under really difficult conditions."

Green's long association with the Shipping Forecast ended when she switched to another BBC newsreading job, but she said she got more letters about the Shipping Forecast than any other work she did on the radio: "I have to say I'm very sorry I no longer read it," she admitted. "I miss it."

Like the Beaufort Scale, the Shipping Forecast changes through time, of course. After its wartime hiatus, the forecast returned in 1949 with the current—mostly—names of the different areas. In 1984, Utsire, off the coast of Norway, was divided into North and South. And in February 2002, what used to be Finisterre (the sea just off the northeastern tip of the Iberian peninsula—the end of the land, hence *finis terre*) got a new designation—Fitzroy—named, of course, for the first head of the Met Office. Fitzroy is sometimes called the first modern meteorologist, and he was the first to apply the term *forecast* to the weather, so he plainly deserves the honor—but it's hard not to consider his suicide and think the honor came just a tad late.

THE BRITISH LOVE AFFAIR WITH THE SHIPPING FORECAST has produced more than a satisfying general awareness of the name Beaufort. Artists hear the Shipping Forecast and they must respond.

Two books—one of photographs, another of watercolors—have been created specifically by artists who based their images on the Shipping Forecast.

For his 1998 book *Rain, Later Good,* watercolorist and author Peter Collyer traveled to all thirty areas of the forecast and made small watercolors of each area, everything from pure seascapes and ship sketches (from places like Sole or Forties, which encompass no land) to beach scenes and lighthouses from those ashore. For Trafalgar (the southernmost area, off the south coast of Spain and even the tiniest northwest corner of Africa), he paints a breathtaking sunset and draws the rear deck of his ship. He writes a bit about each place as well, describing people and trivia—appreciating, for example, nautical language that's worked its way into everyday language (he mentions "down the hatch," though he admits that his own seasickness demonstrated that the hatch allows things to go up as well as down).

But the introductory essay to *The Shipping Forecast,* a book of photographs by Mark Powell, cuts a little closer to that feeling Green was talking about, "intimately linked to a sense of . . . the romantic British Isles." The Shipping Forecast, says writer David Chandler in the essay, is verbal—it's on the radio—so the images are left to the imagination. "In an age dominated by technology and miniaturisation, . . . by the gentle tap of the keyboard and the flicker of the screen, the forecast stirs our residual contact with the sublime," he says. Just as important, says an article from the *Telegraph Magazine,* which Chandler quotes, is the point all my British associates made: The Shipping Forecast intensifies your own comfort, safe in your bed: "With the bed-clothes pulled high and the radio turned low, the promise of a gale at sea is as comforting as the rattle of rain on the window or the howl of the wind in the trees." The Shipping Forecast is sufficiently renowned that you can buy a tea towel with the different regions on it. They made a present of one to me at the Met Office.

BUT AS INSPIRATION FOR ARTWORK, the Beaufort Scale rates more than just its intrinsic inclusion in work about the Shipping Forecast. In the summer of 2002, Irish visual artist Cliona Harmey and electronic musician Dennis McNulty created a piece called "SeaPoint." Inspired by the "poetic language" of the scale, each day for a month they took a ten-second digital video from a bridge on the south Dublin coast, recorded sound as well, and made a series of twenty-eight tiny electronic movies viewable on the Internet (http://rhizome.org/artbase/7110/seapoint.html was the URL at the time of this writing). Each little snippet was titled only by a date and the number and brief description from the Beaufort Scale (the Petersen state-of-sea version): "23.06.02; #3, Large wavelets, crests begin to break, occasional foam." These images weren't the first—a 1971 issue of *Art in America* includes a series of thirteen photos by Jerald Maddox, the curator of photography for the Library of Congress, showing fallen trees, waves, and other results of winds of different strength.

In fact, though, you hardly needed to wait until the 1970s for images illustrating the Beaufort Scale. Don't forget, the first series of images for the wind scale was prepared before Beaufort even entered the narrative—Dalrymple, after all, when he wrote his will in the late 1700s, included his "series of prints . . . to explain the degrees and gradations of wind from a calm to a storm." He also mentioned them in his *Practical Navigation:* "I should despair to explain them, by any description in Words, to Persons without experience. . . . but the Effects are distinctly visible in *Pictures of Sea-Pieces* in my possession." The prints are long dispersed, so which images Dalrymple had in mind nobody knows. The point is, long before artists responded to the scale, using pictures to illustrate the levels of the Beaufort Scale was old news.

Hanging on the wall in the basement of the Met Office archive, in fact, is what is probably the first set of drawings ever undertaken specifically to illuminate the Beaufort Scale. It's a series of cartoons, based on the 1906 scale—both the shore criterion and the specification for coastal use based on the behavior of fishing smacks (3: Smacks begin to careen, and travel about 3–4 miles per hour; 6: Smacks have double reef in mainsail. Care required when fishing).

The drawings are wonderful, largely because they depict the images conjured up, it seems, in everyone who reads the scale. For 0, calm, there is a little sea captain (in sweater and fisherman's cap) rowing a skiff and smoking a pipe. The smoke in the pipe rises vertically. In 1, light air, the captain is joined by another, and they stand by a harbor smoking. The smoke from their pipes drifts perceptibly, though the windvane across the harbor behind them—atop a steeple, just the way one imagined—remains stationary. In 2, the second captain is now piloting a smack, whose sails are filling with the light breeze. The first captain is still on shore, looking through a telescope; that vane is still visible beyond him, though its indicator has pointedly changed directions. In 3, visible in the background, the smack begins to careen; it might be that both captains are aboard, because on shore they've been replaced by two fellows playing golf on a tree-filled course near the harbor, enabling us to see the "leaves in constant motion"; at the hole, where one of the golfers has just landed an approach shot that leaves him a six-inch putt, the pin is still standing, showing that "wind extends light flag."

By force 5, we see "wavelets form on inland waters"—a puddle created on the beach by a little boy with a toy sailboat that mirrors the smack, visible behind him with shortened sail. In 6, a man beneath a pole supporting telegraph wires cups a hand to his ear to hear the whistling, and in 7, a woman struggles against the wind; her umbrella is furled, which is wise—beyond her one flies, inside out, through the air, demonstrating that we've passed force 6, "umbrellas used with difficulty."

It's a thing of beauty and delight, the more so perhaps because nobody knows who made it or why—it's identified by the Met Office only as a print of "one of an extensive collection of late Victorian and Edwardian slides of meteorological interest held by the [Royal Meteorological] Society." Just another anonymous contribution to the Beaufort Scale.

ONCE YOU'RE AT THE MET OFFICE ARCHIVE, it's worth taking a look at what is now a standard issue of the office: a book of photographs of the state of the sea for every step on the Beaufort Scale. Called the *State of Sea Booklet,* it's a straightforward, spiral-bound set of a dozen or so plasticized pages, each showing the front end of a ship and the ocean in one of the states defined by the Petersen state-of-sea scale. It's a truly useful thing, if you're actually on a ship and are trying to gauge the state of the sea for your weather log. It hardly seems likely that a sailor will hold the booklet in his or her hand and then glance up at the sea to compare, but during calmer moments it probably provides a helpful way to get a general sense of what the sea looks like when the wind is at a particular level. And in any case, it surely would have satisfied Dalrymple.

There are plenty of other similar pictures available. If you go to the National Maritime Museum in Greenwich, you can find a display of the Beaufort Scale, showing you simple cartoons: house, tree, flag-pole. In 0, everything is still, and the smoke rises straight out of the chimney. By 5, the smoke is flowing sideways, the tree leaning, the flag snapping full length; by 8, roof tiles fill the air. By 9, the tree is coming to pieces; by 11, the house is; and by 12, there's nothing left. It's a

On a website using stamps to illustrate the Beaufort Scale, this stamp illustrates Beaufort Force 5, fresh breeze: "Moderate waves taking a more pronounced long form; many white horses formed; chance of some spray."
Image from www.seemotive.de, website copyright Bjoern Moritz.

simple series but it gets the job done, it's ubiquitous on the Internet now, and once again Dalrymple can rest easy knowing his illustrations of the state of the wind have found their way into the culture.

Just the same, it seems that Dalrymple wanted to use paintings that already existed to demonstrate the effects of the wind. That is, rather than create images to suit his purpose, he wanted to use images already drawn from observation, organize them, and say, "See? This is what I mean by 'fresh gale.'" The benefit, presumably, would be that the artist would already have observed details that someone trying to draw didactically wouldn't think of. Starting with reality makes a better picture than drawing from concept.

Nicely, Dalrymple's approach is still being undertaken now and then. The BBC—with the Shipping Forecast to their credit, how could they not?—recently created a television series called *Painting the Weather,* in which dozens of paintings from collections throughout Britain were assembled and organized according to weather. You can find pictures about wind, about rain, about fog, snow, the seasons. Because the collection is now on the Internet (http://www.bbc.co.uk/paintingtheweather/), you can organize the pictures by any of ten different weather symbols (fog, gales, windy, hard rain, and so forth). For example, a painting by Fionnula Boyd and Les Evans called *Evidence: Wind* shows a young girl walking away from a painting of a windy scene, her own hair being blown forward by what looks to be a wind of force 3 or so. *Through Wind and Rain,* by William McTaggart, on the other hand, shows a small boat off the coast of Scotland facing what looks to be a wind of force 7, a moderate gale—according to Petersen, "Sea heaps up; white foam begins to be blown in streaks along the direction of the wind." The BBC doesn't include any versions of the Beaufort Scale on its website, but the fact is once you know the Beaufort Scale, you can use it on paintings as easily as you can on the weather outside.

Still, if you ache, as Dalrymple did, for that series of images, made independently, and then organized according to the wind scale so that the scale is illustrated by the discrete details the artists chose independently, you actually have one available to you now.

It's on stamps.

Yes, stamps. The Ships on Stamps unit of the American Topical Association of the American Philatelic Society has on its website (http://baegis.ag.uidaho.edu/%7Emyron/html/wind.htm) a series of twelve stamps from various countries (among them Monaco, the Maldives, Belgium, and Guernsey) depicting ships at sea, going from 0, in which a twopenny stamp from the British Indian Ocean Territory shows the fully rigged barque *Westminster* becalmed in a glassy sea. By force 3 a Liberian steamer plies through large wavelets with a few crests. Force 7 shows a Swedish battleship steaming through the foam; force 10 shows a two-masted steamer from the British Virgin Islands in trouble from "very high waves with overhanging crests"; and the stamp from Monaco, illustrating force 12, shows a desperate group of sailors trying to lower a boat over the side of a square-rigged ship in terrible trouble.

I think wherever Dalrymple, Beaufort, Smeaton, and Petersen are even now playing whist—or, more likely, watching the air move over the Earth and discussing its causes and effects—they raise their glasses to the Ships on Stamps people.

From the children's book *The Rising of the Wind,* Beaufort Force 6: "Umbrellas used with difficulty." This book came about when artist Claire Forgeot drew images of the Beaufort Scale on a dare, and the images were too lovely not to publish.
Reproduced by permission of the artist.

AS FAR AS BEAUFORT SCALE ARTWORK GOES, though, probably the world champion is *The Rising of the Wind: Adventures Along the Beaufort Scale,* a children's book with artwork by French artist Claire Forgeot and text by writer and musician Jacques Yvart. The artwork is the core of the book, resulting from a conversation Forgeot had with her brother-in-law in which she averred that imagination was like a muscle, requiring exercise. As an example she offered to illustrate any subject he chose. A sailor, he offered the Beaufort Scale. The resulting pictures were so gorgeous that Forgeot showed them to her editor, who found Yvart and asked him to write a story to accompany them. The full-page paintings illustrate each level of the scale, showing the land, the sea, and the shore; included in each piece are three descriptions: the one from the 1906 shore criterion; the one from the coastal specification, describing fishing smacks; and the one from the 1927 Petersen state-of-sea scale.

If the cartoon from the Met Office didn't conclusively demonstrate the clarity of those descriptions, *The Rising of the Wind* seals the deal—the images are similar enough that it could be the second edition of the cartoon. In 0, a fisherman smokes a pipe—the smoke rises vertically. By 2, we see a fishing smack with its sails filled—and back on shore, of course, the church, with the steeple, atop which is that "ordinary vane moved by wind." By 4, loose paper is blowing along the shore, and red sailboats heel on blue waves. Force 6 is filled with umbrellas sailing through the air; and by 9, chimney pots and slates are clearly being removed. Those simple Beaufort Scale descriptions are so clear that it seems they create the same images in everyone who reads them. In fact, the book is dedicated to Beaufort, and the preface lauds the original Beaufort Scale, which demonstrates "with its easily graspable descriptive terms, so useful to the layman . . . that there is a poetry in the winds that no mere instrument can measure."

Well, not meteorological instruments, anyhow. Yvart is a musician as well as a writer, and for each level of the scale he wrote a song to accompany its story and picture. Yvart eventually recorded those songs, and he sent me a tape. Played mostly on harp, flute, and guitar, they sound much like sea shanties, starting of course with a quiet, meandering song for force 0 and finishing in a roar by the time they reach hurricane. Yvart, the son and grandson of sailors, lives in Dunkerque, by the sea, and the sea is part of his life, so the songs came naturally to him. What's more, Yvart sent me more than just the official record of the book's songs. Later, after further e-mails, he also sent me a cassette of a live 1997 performance of those songs in front of a highly appreciative Dutch audience. Though Yvart sings most of the songs in French, he sings three in English, and narrates a bit in English as well. He sings one song in Dutch—it gets a big hand—and another in Frisian, a language spoken in the Frisland province of the Netherlands. But, expressing the international spirit of the Beaufort Scale perhaps better than any other scale aficionado, Yvart sings the remaining song in Esperanto.

Surely if there's one person in history who would have embraced Esperanto, it would have been Francis Beaufort, the man who wanted everything communicated so everyone could understand it. It bears repeating: a song about the Beaufort Scale—*sung in Esperanto.* And as songs about the Beaufort Scale in five languages demonstrate, the Beaufort Scale has become international in ways for which even Beaufort could scarcely have hoped.

And, of course, Yvart wasn't the only musician to fall for the Beaufort Scale. In 1984, the Finnish composer Aulis Sallinen, something of a national musical hero in Finland, created something that, if it were better known, might gain him similar status in the English-speaking world. He wrote, for unaccompanied mixed choir, his Opus 56—a work called "The Beaufort Scale."

It's exactly what you'd expect. Six minutes long, with an almost whispered beginning ("Calm . . . smoke rises vertically . . . sea like a mirror"), rising in pitch and intensity through all 12 numbers of the scale, finishing at "hurricane" with the choir murmuring the Morse code SOS pattern. Sallinen takes great pleasure in some of the nicer turns of phrase in the scale—in force 1, the choir lavishes attention on the phrase "direction of wind shown by smoke drift," repeating it several times, but then, in quick staccato, adding, "but not by ordinary windvanes." There's a lot of counting and a fair amount of imitation of the wind. It's lovely and brief and in all, if you were going to sing the Beaufort Scale, Sallinen figured out just how you'd do it.

BUT SALLINEN'S HUMORESQUE BROUGHT me back finally, to where I'd started with the Beaufort Scale. Sallinen set those words to music not because they were useful or fascinating or historically interesting, though they manifestly are; Sallinen set them to music because they are beautiful—because they are, fundamentally, poetic. And that brought up something I couldn't avoid recognizing about Beaufort.

The more I learned about Francis Beaufort, through his biographies, through his letters and records, and through research into hydrography and the Admiralty Chart, the sailing ship and the windmill, the more admiration I had for him, yet the clearer one thing became: he was no poet. This was a guy, after all, who was shot in the groin by a gang of armed Turks, shot badly enough that he recovered only by miracle and was retired from active duty as a result. Still he retained command of his little boat, and he describes it thus: "Before I fainted from the loss of blood, I had the satisfaction of sending her round to rescue the scattered officers." The officer who took command when Beaufort passed out, Beaufort tells us later, only "with difficulty, indeed, . . . could curb the natural fury of the boats' crews, which, if unrestrained, would speedily have taught those miscreants a wholesome lesson of retaliation." This is many things—it's the way people wrote for the public in the early nineteenth century; it's the tone of a man full of himself (Beaufort was at least forty when he wrote that, not the young rakehell he sounds like); it's an example of everything the Beaufort Scale is not—but it's not poetry.

The more I learned about Dalrymple and Smeaton, about Tycho and the myriad sea captains and weather observers who tried to organize their perceptions of the wind, the more I drew the same inescapable conclusion: Whatever these guys were, poets they were not.

And yet the Beaufort Scale as I found it, in its 1906 glory, is nothing if not poetry. It has those iambs and trochees, in tetrameter, pentameter, over and over, and all accidentally.

But of such accidents is genius composed. Of Satchel Paige's remarkable list "How to Stay Young" (the final suggestion is his famous "Don't look back—something might be gaining on you"), I wondered for a long time why of the six items the one I instantly memorized was the first: "Avoid fried meats, which angry up the blood." It's funny, and folksy, but I wondered about it until I noticed it was iambic pentameter, the language of Shakespeare and the poets. I had it memorized before I knew it—it's self-memorizing. It's possible that the way separating, classifying, organizing is a fundamental human thought pattern, iambic pentameter is a fundamental speech pattern, at least for speakers of English. In any case, it surely works. "Nature, rightly questioned, never lies." That's the wonderful, epoch-defining quote from the *Manual of Scientific Enquiry,* and it's nine syllables long. Drop in an elided "but" or "and" before it, and you've got iambic pentameter; leave it as it is and finish with a one-beat pause before the next phrase, and you've got trochaic pentameter.

Even the Shipping Forecast, which contents itself with the mere numbers of the Beaufort Scale, has the same feel. "It was very akin to reading a five-minute poem," newsreader Charlotte Green told me. "It was lyrical, with internal rhythms and cadences."

So Beaufort wasn't a poet; Sir George Simpson, the engineer from whose desk the Shore Criterion emerged, was no poet. Dalrymple, Smeaton, Tycho, on and on—no poets these. Yet that 1906 version of the scale, the Shore Criterion, is poetry if anything is. And if artists noticed and musicians noticed, you might bet that poets did too, and you'd be right.

In year 1979, Seamus Heaney wrote a sonnet using as its inspiration the Shipping Forecast ("Dogger, Rockall, Malin, Irish Sea: / green, swift upsurges, North Atlantic flux / conjured by that strong gale-warning voice . . ."), but that's the Shipping Forecast, not purely the Beaufort Scale. Not long after my obsession with Beaufort's scale became known, a friend pointed me in the direction of a poem called "The Scale of Intensity," by Scottish poet Don Paterson, from a 1997 collection.

"Not felt," the poem starts. "Smoke still rises vertically. In sensitive individuals, déjà vu, mild amnesia. Sea like a mirror." A few lines down the intensity has progressed: "Small bells ring. Small increase in surface tension and viscosity of certain liquids. Domestic violence. Furniture overturned." Not much further along you find "Perceptible increase in weight of stationary objects: books, cups, pens heavy to lift. Fall of stucco and some masonry. Systematic rape of women and young girls. Sand craters. Cracks in wet ground." By the time Paterson's "Scale of Intensity" reaches its conclusion, you find "Standing impossible. Widespread self-mutilation. . . . Most bridges destroyed. / Damage total. Movement of hour hand perceptible. Large rock masses displaced. Sea white."

In an e-mail correspondence with me, Paterson said he first encountered the Beaufort Scale—the 1906 Shore Criterion—in school. Only later did he find the 1927 Petersen state-of-sea scale, after which he went looking for similar scales and found the Mercalli scale and others, whose language he appropriated, adding his own emotionally flat descriptions of the horrors of war. "The point was, I guess, to make a wholly neutral poem about war," he said. "It was written during the Bosnian crisis. I was appalled by the sentimentality of some of the poems being written then—as if we needed to be told how awful it was, as if an emotional response to these events needed to be *prompted*."

Though he emphasized the scale's dispassionate tone, that wasn't what first drew him to it, he said: "The language is nonetheless intensely poetic, since it marks the points on the scale by one well-chosen specificity. That's so often what poets are trying to do—find the one tiny thing that stands for the larger thing." In addition to discussing his own poem, Paterson did what poets always do, which is point me to another poem about the Beaufort Scale, this one by the American Paul Violi.

Violi seemed amused that Paterson had sent me his way, since "the poem . . . has never been published." The poem he wrote, he told me, wandered off from the Beaufort Scale and became "a satiric muddle" that he occasionally compliments himself for deciding not to publish. Just the same, the Beaufort Scale still reaches out to him. "The scale seemed to be a poem already, almost 'a ready-made,' in that instead of relying on numbers for measurement, it offers terse, sharp images in a dramatic movement, moving from the objective to the subjective, from the precise to the broad, as the force it describes increases. Also, as poems on some level do, it attempts to catch something by describing its effects." And succeeds rather remarkably, of course—in any case, well enough that a poet like Violi found it so satisfactory that he chose never to publish his poem about it.

But the poet I found who seemed to be most closely following the dictates of the Beaufort Scale was Canadian poet Nelson Ball, who I think is not merely celebrating but applying the scale. In 1967, Ball published a collection of poems called *Beaufort's Scale*, which he wrote after accidentally discovering the scale, to his delight, in the Merriam-Webster *New Collegiate Dictionary*. He printed the dictionary version of the 1906 scale on the back of that book (it's from the Sixth *New Collegiate;* you can tell because it still has the map symbols), and though none of his poems directly addresses the scale, his approach is plainly colored by the scale. "I thought the 'descriptions' of the visual manifestation of the various wind strengths were wonderful—simple, plain, direct—almost like imagist poems," he told me in a letter. "For this reason I chose the title for my book." His poems are brief and imagistic: "Caught between movements / of air / the tree / is / almost defined—form / imposed as if there were no wind. / The wind / changes—a new / definition," one runs in its entirety.

And his work has maintained its Beaufortian direction. In his most recent book, the 1999 *Almost Spring*, his poem "In the Carolinian Zone" could be a kind of scale itself. It runs as follows: "we measure / snowfall / by the visibility / of corn / stubble." A summation of the

Nelson Ball Scale of Snow Depth there, in nine words. "Signs" is even more so: "Snow / thick / on / evergreens / begins / to crack— / wind / rising / Wind- / blown / snow / adheres to / tree- / trunks." That would be about a 5 on a winter version of the Beaufort Scale. I'm surprised someone hasn't written one yet.

I LOVED FINDING OUT ABOUT THESE POEMS, reading them, speaking with the poets themselves. It deeply satisfied me to find that poets—and artists and musicians—had been fascinated by the Beaufort Scale since its inception. Or, more accurately, since the writing of that 1906 scale. That's the one I found—as did Nelson Ball—in the dictionary. It's the one that helped inspire the poets, that inspired Claire Forgeot and Jacques Yvart. In Sallinen's humoresque, at "calm" he includes the line "sea like a mirror" from the Petersen state-of-sea scale—after that, every word of the scale comes directly from the 1906 Shore Criterion.

That's the poetry. That's what inspired the Petersen version after all, and it's what lay in the dictionary, waiting, for all those years. But though I had found poets who had written *about* the Beaufort Scale, I had still never really found the poet *of* the Beaufort Scale. There was a poet in there, I was just certain of it, and I was determined to find him. So I went back one last time to my tattered photocopy of that 1906 Publication 180, looking for clues.

The highlight of the publication is the lovely two-page table of the final scale—it includes six different expressions of the Beaufort Scale, all lined up together, then followed by seven different corresponding numerical measurements (mean velocity of wind in gusts or in lulls; the maximum and minimum velocity of each Beaufort number, and on and on). Beyond this, though, the publication concerns itself almost exclusively with finding velocity equivalents, accommodating for a staggering array of complicating factors: the height above the ground of the anemometer; the wind direction; the number of times different strengths of wind were observed at different stations. Sir Napier Shaw (director of the Met Office), Sir George Simpson, and the rest of their helpers basically conclude that it's all guesswork, though they do their best to come up with constants as useful as possible.

But naturally they also, in great detail, discuss their methodology—and there I found one last chance at the poet of the Beaufort Scale. There I encountered the North Shields observer.

THE NORTH SHIELDS OBSERVER was one of five observers at five different daily weather stations around England—at Scilly in the southwest, Holyhead on the northwest Welsh coast, Yarmouth on the east coast, North Shields on the northeast coast, and Oxford in the middle of the island. From 1900 to 1902 the daily estimates based on observation from those stations were organized and compared with anemometer readings from the same stations. The observers on the sea would estimate the wind by what they saw in their harbors and on land. The Scilly and Yarmouth observers were "men who have been connected with the sea all their lives," and who both also had excellent views of the sea from their places of work. Their observations covered almost exclusively the behavior of fishing smacks. The observer at Holyhead was far enough from the harbor that though he too described boat behavior, he based his observations largely on the coast guard flag at his station. The Oxford observers did nothing more, it turned out, than count revolutions of anemometer cups, compare the speeds with a chart they got out of a book, and report those numbers to the Met Office: "When asked to describe the effects of the different forces upon the trees, smoke and people, they were not able to do so," says Publication 180.

That leaves the North Shields observer.

In a table, Publication 180 lists observation summaries prepared by each observer. The Scilly and Yarmouth observers talk about what sails ships have up, occasionally throwing in notice of other details: "White horses with E. breeze," says the Yarmouth observer of force 4; "Boat begins to throw water, so that men must put on oilskins," says the Scilly observer of force 5. The Holyhead observer says things like "Flag out straight, but droops," of force 4, or of force 7, charmingly, "Flag is generally taken in by coast guard."

The North Shields observer, meanwhile, says things like "Slight breeze can be felt on face. Smoke traveling slowly," of force 2, or of force 5, "Umbrella easily managed, but hat must be held at street-corners."

"Hat must be held at streetcorners." Or, in other words, *That's the guy.*

That's my poet right there. It was the work of only a few minutes in the Met Office archive to find the original weather reports for North Shields from those years. It was a good period: satisfying brown shiny paper printed forms that folded up to become envelopes for mailing. You can even see the stamps change, in 1902, from Victoria to Edward. More important, they identify the observer as a certain George Clark, head clerk of the post office, which fit just fine with Publication 180, in which Sir George Simpson describes him as "employed at the post office, while the thermometer screen is situated in a square surrounded by houses about five minutes' walk away."

George Clark's five-minute walk, I concluded, was the core of the Beaufort Scale I had found all that time ago in the dictionary. "When walking from the office to his screen, the observer passes through narrow streets, and forms an estimate of the wind from its effect upon the dust, smoke and people," Sir George writes of the North Shields observer, "and when he arrives at the square looks at the cups to see if they are revolving as he would expect from his estimate." The result was the series of phrases I knew: "Leaves in motion"; "Raises loose paper and dust"; "Trees shaking and wires whistling"; "Whole trees moving." Without a doubt, I had found my poet.

NORTH SHIELDS IS A LITTLE FISHING TOWN at the mouth of the River Tyne in northeast England, close enough to Newcastle to be on its subway system. Row houses with tile roofs march in ranks away from the center of town, descending the hill toward the river. There are two lighthouses—the Low Light and the High Light—down by the fish quay, where boats still unload catches of haddock and prawns. In the center of the brown-and-gray Victorian town is Northumberland Square, a small "square surrounded by houses." In the library I found maps from 1901, and on the maps I found the old post office, now boarded and empty. Just as Sir George said, it was about five minutes' walk away.

So I made that walk. Along Saville Street, in front of brownstone and brick houses and stores from the nineteenth century. An old fire station that the observer would have passed is now the Bell and Bucket pub (the people there are very friendly), and Norfolk Street leading to Northumberland Square is notable for more of those tile roofs, and for the chimney pots that extend above them. Northumberland Square is a neat green square surrounded by wrought-iron fence; it has some lovely trees, a few benches, and, in the middle, a wooden statue of a fishwife called by locals "the wooden dolly," the fifth of her tribe. The first, probably originally a ship's figurehead, appeared in North Shields in the mid-1800s, and it became a tradition for fishermen to whittle off a piece of the wooden dolly on their way out to sea for luck—hence she was always diminishing and periodically needed to be replaced. One of her ancestors would have stood in North Shields as Mr. Clark made his daily weather walks.

There's much less wind in the square, which is sheltered by houses, than by the river. I took a weather note there ("Northumberland Square: Flagpole cords flutter but don't slap against pole; outside boughs of lighter trees wave; seagulls noticeably breast wind and pick up speed wheeling away from it") and stood for a moment. So this was where the North Shields observer had looked around, collected his thoughts, and strung together the lines of description that resonated unceasingly in my mind. A nice little park. That seemed to be it, and I was disappointed.

Because here's the thing. As I wandered around in the physics of wind and the history of its measurement, I eventually began to feel that I got to know the players. Beaufort, the guy who could do everything, was always paying attention, and would help you in any way he could—the world's best and coolest older cousin. Smeaton, the logical, even-tempered guy, who tried to build every single machine in the world in his garage—and they all worked; Dalrymple, the perpetually annoyed geezer, withholding charts from the Admiralty Lords because they hadn't let him go on a boat ride in 1765. It's not as if I truly understood them, but I felt that I had gotten to know them a bit.

Naturally, once I identified him, I wanted to put a similar face on the North Shields observer. So I went looking for him.

THE 1901 CENSUS (it's online) told me where he lived in 1901, as well as the names of his wife, three kids, and nineteen-year-old Norwegian au pair. The North Shields Public Library yielded church and parliamentary records that supplied the date and place of his 1891 wedding and another house he lived in. And his middle name, Bush.

I conceived a plan. Someone capable of those straightforward, almost surgically precise descriptions, I thought, could never have limited himself to meteorological observations. The North Shields observer would have had children, and his children would have done the same. Somewhere, I was sure, in some attic or basement, one of those descendants had a package of envelopes wrapped up in a ribbon, and they were the letters of the North Shields observer to Margaret, his wife; or a pile of bound leather notebooks that were his journals; or . . . something. For my final pursuit of the Beaufort Scale, I resolved to find those.

It was a pretty stupid idea, though in retrospect I did better than I should have expected. I went to the public records office in London and began running my finger down lists of Clarks in the big green marriage registers. Amazingly, I did find Isabella Clark, daughter of George and Margaret. Isabella married Mr. George E. Betts in the spring of 1914 in North Shields. I never found another mention of either of the other children of the North Shields observer (William A., age six in 1901, might have been the perfect age to die in World War I; his sister Elmar, three in 1901, would have been the perfect age to, as a result of the war, find no husband). Isabella, though, according to birth records in the big red books across the room, was fecund and had George Edward Betts in November 1914 (do the subtraction problem and—Isabella!). Patient scrolling through the years starting when George E. Betts would have been sixteen or so finally yielded, if it's the same guy, the marriage of a George E. Betts in 1953 to a woman named either Mona Alexander or Mona Hall (the marriage records are unclear, listing her both ways), and in 1964, the red books say, was born Edward P. Betts, to a mother whose maiden name was Hall. No other children that I could find.

Which meant that as far as I could tell, one Edward P. Betts was my single shot at descendants of the North Shields observer. The Internet, you probably know, is worse than useless when seeking telephone numbers. But as plan B, there are 154 telephone directories that together cover the British Isles. If you sit down in the North Shields public library for a few hours, you can go through every single one of them.

Not one lists so much as an E. P. Betts, much less an Edward P. Betts.

STILL GAME, HOWEVER, I BEGAN CALLING THE forty-one E. Bettses I had found by combing through the combined phone directories of the United Kingdom. I never found Edward—though I did ring a surprisingly large number of old women who, seemingly presuming that I was either a masher or a telemarketer, proved themselves, if not poetic, at least highly creative linguistically.

It's probably just as well. I had a couple of pleasant conversations with people who seemed to hope that I eventually did find the descendant of some guy, which is as clearly as I think I was ever able to explain what I was looking for. But what would I have said if I actually had found Edward P. Betts, if he even was the descendant of the North Shields observer? "Your great-grandfather was rather observant," I would have said. "He had a nice way of putting things, and a hundred years later I got to wondering whether I might paw through your attic." I didn't think that would make much sense to anyone else. Nonetheless, grasping at straws, I went so far as to check the postal pension records for George Clark, and I learned that he had missed exactly three days of work in his last five years and ended up postmaster in a place in Northumberland called Keighley. And that was it.

In a way it was enough. The one thing I had learned about his character—he almost never missed work—fit in perfectly with the exacting quality of his observations, and with the requirements of a weather observer, who, after all, had to be on the job every day for his observations to have value. Beaufort himself had never missed work and had recorded his weather observations every day. The North Shields observer, the man whose crisp turns of phrase were at the core of what is best about the Shore Criterion of the Beaufort Scale, seemed finally, and appropriately, to be just another in the line of borderline anonymous contributors the wind scale seems to have drawn to itself since its first scattered creation in 1582 by Tycho. I would have liked to find out more about Mr. Clark, but I hoped he—and Mr. Betts, the possibly still-living person who is possibly his great-grandson—wouldn't mind. Then again, if the North Shields observer left nothing behind him other than his contribution to the wind scale, he's part of a long line of patient and attentive people—a line running from Tycho through John Smith and Daniel Defoe, through Smeaton and Dalrymple and Beaufort himself, and moving forward even now, with the contributions made by the mysterious figure who decided that "moving cars veer" at force 8. He's part of the club—so are Friendly and Courtney, Beaufort's biographers—and Cook, the Dalrymple guy, and for that matter everyone who lies in bed at night and listens to the Beaufort Scale numbers in the Shipping Forecast. We're all paying attention. We're all keeping the Beaufort Scale alive. We're all North Shields observers.

CHAPTER 9

Observation, a Panegyric: On the Beaufort Moment

Beaufort labeled this drawing "The Oak 50 yards E. of Edgw. door Oct. 6 1830." He made the drawing purely for pleasure, but it almost provides for artists the same model of simplicity and clarity the scale does for writers.

MY HIGH SCHOOL physics teacher once assigned our class a fairly simple problem. He reminded us of certain facts: The Earth orbits the sun once in about 365 days, and the planet meanwhile spins every twenty-four hours. The moon circles the earth once in about twenty-eight days, revolving on its axis exactly one time per orbit. Given all that, he asked us to answer a simple question: From a given vantage point on Earth, does the moon appear to rise earlier or later on successive days?

We all got busy—365 divided by 28, no wait . . . the earth's surface rotates in an easterly direction, and the moon travels . . . no, let's see, if there's twenty-four hours in a day and twenty-eight days in a . . .

After only a moment, the teacher slammed a yardstick on the black laboratory bench up front and instantly got our attention. "No!" he said, probably the only time I ever heard him raise his voice. "You should not be calculating this," he said, looking at us with what I think was sadness. "This is something you should *know.* The moon comes up every day. Every day of your lives. You should be able to answer this question without thinking because you should know when the moon comes up. You should be paying attention." We were very quiet.

I can tell you now that the moon comes up between twenty minutes and an hour and a quarter later every successive day; if you estimate a little less than an hour later per day, you'll be in the ballpark.

That is a Beaufort moment. And that is what I finally figured out the Beaufort Scale was trying to tell me. The Beaufort Scale is about paying attention. It's about noticing whether smoke rises vertically or drifts, whether it's the leaves shaking or the whole branches, whether your umbrella turns inside out or just rattles around some. More, it's about taking note of those details, filing them away, in memory or, as the *Manual of Scientific Enquiry* would have it, in the notebook you'd never leave the house without (along, of course, with your pencil and your map and compass). It's about being able to express what you've seen simply and clearly, in as few words as possible, so that others can share it. It's about the good of sharing that knowledge, of everyone paying attention so that, together, we can all learn as much as possible.

The Beaufort Scale is a manual, a guide for living. It's like a cross between the *Boy Scout Handbook* and the *Old Farmer's Almanac:* a bunch of cool information that you'll never be sorry you have, and a general policy of being prepared to deal with it: to notice that information and share it for the good of everybody involved. It's a philosophy of attentiveness, a religion based on observation: an entire ethos in 110 words. One hundred ten words, that is, and four centuries of backstory.

I BEGAN RECOGNIZING WHAT I CAME to call Beaufort moments not long after I received the little handheld anemometer I bought off the Internet. I received it a couple of days before my forty-third birthday, and it seemed auspicious to begin my wind diary then and plan to keep it up every day for a year. It's possible I kept it up every day for ten days, but I know I didn't make two weeks. Nonetheless, I liked having the anemometer around, though the greatest thing it taught me was that I didn't need it. One day I pulled into a gas station and the wind was blowing so hard I could barely open the door of my car. That seemed worth noting, so I patted my pockets—and was disappointed to find that for some reason I did not have the anemometer with me. I was about to give up on making a wind notation—and then, of course, I realized that I was exactly missing the point: It was precisely for moments like this, when no instrument was available, that the Beaufort Scale was designed. I made my notation, but that was key. Using my own eyes, my hands, my car door to observe the wind, to make note of it, and the hell with the anemometer—understanding the wind by experiencing the wind—that was a Beaufort moment.

A TELEVISION AD I CAN'T AVOID RECENTLY urges people to buy something through the example of a woman portrayed rushing through the streets of a big city, focused entirely on the movements of her watch, desperately trying to make an appointment. As she races around she is so focused on time that she ignores naked men, costumed chorus girls, and zoo animals who are wandering the streets. The message: This woman is wasting her life focusing on her watch, unaware of the pageantry that surrounds her if she'd only lift her head. The commercial advertises a cell phone that can take pictures. Left to yourself, it implies, you will ignore the world. Even worse, it claims, if you have the option merely to perceive the world, you won't take the time. Why would you?

On the other hand, if you have a nice new piece of technology that enables you to digitize your perceptions and bombard your acquaintances with them, you will sit up and take notice.

Welcome to hell.

That commercial to me is the anti–Beaufort moment—it's the exact opposite of everything Sir Francis Beaufort, the *Manual of Scientific Enquiry,* and all the subsequent observational scales have stood for. Everything I have learned about the Beaufort Scale, about the ones that preceded and followed it, and about the people involved with them, demonstrates that our own body is the greatest perceptive instrument ever designed. Not only can it perceive sound, light, movement, temperature, aroma, taste, and time, but it possesses all the processing capacity and speed necessary to organize, categorize, and express that mass of data almost instantly, to then retain and reorganize it, and to find in it utility and value and meaning. With neither cell phone nor LCD screen, neither keyboard nor electronics, Beaufort interacted with things, not representations of things. Beaufort lived an *un-virtual* life, and the Beaufort Scale embodies that spirit.

And then some boob with a new gewgaw to sell tries to convince you that not only is your own perception insufficient to entertain even yourself, but that if you're inattentive to the world the problem isn't you—the problem is your technology. In a magazine for weather enthusiasts I found an advertisement for something called the Thunderbolt Storm Detector, which "tracks storms and calculates

ETA from 60 miles," keeping you apprised of your impending doom with the obligatory flashing LEDs and even an audible alarm. Which means that even the people most interested in weather—the people who have learned to read the clouds and smell the wind, who know the weather signs and have, through long observation, acclimated themselves to the actual behavior of their atmosphere—are still liable to just forget the whole thing and let something with batteries do all the work. These are *weather people.*

In Flamsteed House, the museum of the old Royal Observatory that is attached to the National Maritime Museum in Greenwich, you can follow the development of the chronometer—from the astronomical almanacs by which ships used the motion of Jupiter's moons to determine their position, through the lovely clocks developed by Harrison until, finally, you come to the modern version: a thick plastic rectangle that gets its information from satellites and looks like a walkie-talkie. If that doesn't drive home the point, on the other side of the room you can do the same with clocks—from the first pendulum clocks of the 1650s all the way through the most modern electronic instruments, which the explanatory text accompanies thus: "With the invention of quartz crystal and atomic clocks, world time is now determined without reference to the movement of the earth or stars." This is presented as a good thing, but it raises a powerful question: Time has been broken up into days and nights, months and years, by the natural cycles of the universe itself for as long as we've known about it. That is, for us the movement of the Earth and stars is the very essence of time—it's good to be able to measure it more accurately, but we divorce time from the movement of the Earth and stars at our peril. And just so with the wind—anemometers are nice, but the Beaufort Scale is a way to keep ourselves tethered to the world, to the wind that does more than abstractly spin anemometer cups: it rocks trees and causes waves, spins windmills, pushes ships.

Beaufort lived in a Newtonian world where stuff happened because of things you could perceive, and where if you got interested and wanted to, you could do something yourself to better understand it.

And if I learned one thing during my search for the meaning of the Beaufort Scale, it's this: So can we, if we only want to. We also live in that Newtonian world, and everything about it is still available to us. The problem isn't science; the problem isn't the media or technology or the Weather Channel or the Storm Detector. The problem is us. We just tend to ignore that connection. And the Beaufort Scale reminds us of it.

IN CHARLES FRAZIER'S NOVEL *COLD MOUNTAIN,* Ada, the woman left alone by the war and forced to learn how to live, is taught a lesson by her new companion, Ruby.

> —What do you hear? Ruby said.
>
> Ada heard the sound of wind in the trees, the dry rattle of their late leaves. She said as much.
>
> —Trees, Ruby said contemptuously, as if she had expected just such a foolish answer. Just general trees is all? You've got a long way to go.
>
> She removed her hands and took her seat again and said nothing more on the topic, leaving Ada to conclude that what she meant was that this is a particular world. Until Ada could listen and at the bare minimum tell the sound of poplar from oak at this time of year when it is easiest to do, she had not even started to know the place.

That's a Beaufort moment. In the Jack London story "To Build a Fire," the unnamed narrator has this experience: "He spat again. And again, in the air, before it could fall to the snow, the spittle crackled.

He knew that at fifty below spittle crackled on the snow, but this spittle had crackled in the air. Undoubtedly it was colder than fifty below." That's a Beaufort moment. Huck Finn, looking for a way to escape his Pap, leaves the cabin one morning. "I cleared out up the river-bank. I noticed some pieces of limbs and such things floating down, and a sprinkling of bark; so I knowed the river had begun to rise." That's a Beaufort moment, too. A Beaufort moment is any moment where instead of merely passing through my surroundings I notice them, notice them in a way that engenders understanding. London's narrator didn't just say, "How about that," he thought, "Colder than fifty below." That's data, that's information; Huck, too, found not just detail but information in his observations.

That kind of observation—useful observation—is what the Beaufort Scale turns out to represent. That's what drew me to it. That's what it has to offer.

On my porch I can hear the whispering of trees as the wind comes up my street. If I can't see the leaves and observe the direction in which they move, or which trees begin to blow first, I can hear it. First the shaggy maple on the corner lot whose owners have the statue of a naked lady on their porch; then, one house closer, the tall maple up the hill next to our building; then the pine across the street. The hickory that stands directly in front of our porch—you can just not quite touch its closest leaves—never seems to make a noise; I think it's too close to our building, and the wind is too much deflected.

If the wind is a little higher aloft, it sighs only in the highest spreading canopy of a gigantic catalpa one house farther along. That one sways in a stately way—its top is so rounded and smooth it looks like the tree at the end of *Go Dog Go,* where all the dogs drive at the end for the big dog party. I can't tell yet the individual voice of each tree, but I'm listening for it.

In the end it's distinctions like these that the Beaufort Scale is about—it's about listening closely enough to make them. And I'm saying much more than that I would like to be able to make those nice distinctions. I think making those distinctions is a fundamentally better way to live. I think when I'm paying that much attention I'm living a better life. It kills me how often I fail to. And the Beaufort Scale puts me in mind to do it more.

I STARTED TO HAVE ANOTHER KIND OF BEAUFORT MOMENT, too, or began noticing that I was, anyhow; the one I remember best occurred along the Thames at the little town of Gravesend. There in 1871 a tiny church was dedicated, by one of his daughters, to the memory of Sir Francis Beaufort. Called St. Andrew's Waterside (it sits on pilings jutting directly into the Thames), the church was built to minister to the people living and working aboard the boats that plied the river. The ceiling inside is pitched and arced like the inside of an upside-down boat. The church has long since been converted into an arts center, but with some friends and their two-year-old son I went to visit.

There wasn't much to see, but there was a pub nearby, so we got a drink and sat at a picnic table to watch the river and play. Young Matthew, however, was cranky and finally sleepy, so instead of playing with him I pulled out my little pocket kite and played with the wind. The wind there shot straight down the Thames; the kite would extend its string, but it never went too high. Surprised to see a kite come out of my pocket, my friends joked that I was like a tool chest, or a Swiss Army knife. I couldn't have been more flattered—it was exactly how I felt, and I thought Sir Francis, the force behind the *Manual of Scientific Enquiry,* would have been proud. By that point I always carried in my pocket paper and pencil, as well of course as a compass, just as the *Manual* bid me. Because I was traveling I kept a map with me too, and usually the pocket anemometer. Why not a kite too? Sir Francis, I thought, would have had one.

When I found, early in my pursuit of the story of the Beaufort Scale, that Sir Francis Beaufort had not really written the scale, I reassessed my goals. Beaufort himself, then, I thought, would get a smaller portion of my attention. It was the scale, with its vivid language and powerful utility, that I was interested in, and Beaufort, I figured, was a mere stopping point, a man with whom the energy of this nascent scale, developing through time, stopped, adopted a name, and achieved the kind of currency that allowed it to become an international standard. Beaufort was just a way station in the journey of the scale.

But I could not have been more wrong about Sir Francis. True, Beaufort was not the author of the scale, but he was never a mere stopping point, either. In every way, what I found marvelous about the scale found a mirror in Beaufort, and what I found to admire in Beaufort found expression in the scale that bears his name. There is a kind of cosmic power in the universe that confers on certain things their appropriate titles, and I believe the Beaufort Scale has found just the name it needs. Beaufort, after all, was a stickler for finding the words that average people used for their mountains, rivers, bays, and shores; he adjured his hydrographers never to slap a name on a feature out of vanity or favor, but to call the places they visited and observed by the names they already carried. Beaufort would have said this was simply a way to improve accuracy, but I'm not so sure. It was also, I think, a way of showing respect for the observable universe, a sense that names have meaning—they find their way onto things through processes long and difficult to understand, and it's best not to mess with them. Just so with the Beaufort Scale.

It's a scale of vivid description that carries the name of a man who was mad for vivid and exact description. It's a scale based on observation given the name of a man for whom observation was something he simply could not stop doing, something that he continued doing when all that was left for him to observe was his own declining health. His biographers point out that Beaufort kept up his weather notebooks and pocket diaries until days before his death; perhaps we should say rather that Francis Beaufort stopped observing, stopped recording, and then he died. He couldn't live without it. No scientist himself, Beaufort nonetheless saw the benefit that could accrue by keeping mountains of data and assuming they would eventually be useful. Thus it is no surprise that the data gathered on the sailing ships, for as long as ships' logs have been kept, are now being vetted and compiled by climatological groups to learn about weather and wind during periods when no other records were kept. Even ships that maintained no wind records have a log of their sail changes—and those changes, through Beaufort's state-of-sail wind scale, can now be translated into specific wind speeds. The *Europa,* the ship on which I sailed, regularly contributed to those efforts; as a ship with the full complement of sails from Beaufort's time, it adds wind data to the records it keeps of its own sail changes, and those records help scientists interpret wind data recorded by ships long before it was even possible to measure the wind. Beaufort turned the sailing ship into an anemometer, and even though most of the ships are gone, their records, like Beaufort's scale, still do work.

This would delight Beaufort. The Petersen state-of-sea version of the scale he propagated is still very much a working scale, used aboard ships without anemometers daily all over the world as they file telegraphic weather reports to national and international agencies, to say nothing of the Shipping Forecast. That would please him too.

But Beaufort himself also emerged as the representative of another element of the scale. I remember opening a box of Beaufort's records in the Huntington Library and finding a long row of his pocket notebooks, lined up like soldiers, one per year, for year after year of his life. And I thought immediately of the row that stands on my own shelf at home, notebook after notebook, and of the similar rows on the shelves of like-minded observation junkies everywhere. I once met at a dinner party a copy editor from the *Washington Post* who carried in his back pocket the exact style of notebook I carry in mine, and we were like lost brothers, like Masons attracted by one another's signets. We talked about what we jotted down and how long it took us to fill a book, but in reality we just enjoyed being near one another, enjoyed finding someone else who knew the great secret, the secret written in the *Manual for Scientific Enquiry:* Never leave the ship without pencil and paper; keep a compass handy; jot down what you see. Keep your eyes open.

We were acolytes in the Church of Jotting Things Down—we kept notes, paid attention, because . . . well, because you never know. We represented the modern descendants of those eighteenth-century contributors to the *Philosophical Transactions.* Sir Joseph Banks, president of the Royal Society, saw one John Cullum in London in 1783, and Cullum told him there had been a frost in his neighborhood in June. "You expressed a desire of receiving some particulars of it," Cullum wrote the following November. "I therefore now send you some memorandums which I made at the time." That's why—you write things down because later someone may want to know. It's just the right way to be.

AND EVERYTHING I FOUND IN MY SEARCH for the story of the Beaufort Scale seemed to lead me down the same virtuous path. Decades ago I stumbled onto this lovely piece of writing, then set out—never in a terrible hurry—to find its creator so I could read anything else he might have written. A modest goal, to be sure. I ended up finding, bit by bit—from Beaufort, Dalrymple, and Smeaton; from the Society for the Diffusion of Useful Knowledge and the *Manual of Scientific Enquiry;* from the Petersen Scale, the updated Savannah scale, the musical renderings of the Beaufort Scale—not just an example of brilliant writing but an entire way of looking at the world.

Not a new way—just a better way. And everything pointed to it. When, emulating Beaufort, I learned to draw, I was amazed by the way even my ham-handed sketch of Montevideo turned out to be more useful to me than the photographs I took—and then I found a comment by John Ruskin, pointing out that I was just learning something that observant people have long known. Ruskin wrote that "photographs . . . are invaluable for record of some kinds of facts," but are no match for the understanding that comes from the observation and effort of drawing. "When once you have paid this price, you will not care for photographs of landscape," he said. "They are not true, though they seem so." Just the same—the TV weather maps boil and swirl like the light show at a rock concert, giving you forecasts and dew point and wind chill and "real feel." It's lovely and it feels like information, but then someone asks what the weather will be, and instead of walking outdoors, we log on to weather.com or we turn on the TV.

I ORIGINALLY FELL IN LOVE with the plain language of the Beaufort Scale in 1984, and then, in seeking its history in the Royal Society, I found Thomas Sprat, in his *History of the Royal Society* (written in 1667), elegantly expressing what I would eventually learn. In a section titled "Their Manner of Discourse," Sprat describes the flabby, complex, self-aggrandizing speech one expected of Natural Philosophers of the time, and favorably compares with it the simpler language used by the scientists of the Royal Society: "They have therefore been most rigorous in putting in execution, the only Remedy, that can be found for this *Extravagance:* and that has been, a constant Resolution, to reject all the amplifications, digressions, and swellings of style; to return back to the primitive purity, and shortness, when men deliver'd so many *things*, almost in an equal number of *words*. They have exacted from all their members, a close, naked, natural way of speaking, positive expressions, clear senses; a native easiness; bringing all things as near the Mathematical plainness as they can: and preferring the language of Artizans, Countrymen, and Merchants, before that, of Wits, or Scholars."

A close, naked, natural way of speaking; primitive purity. Positive expressions, clear senses. The Beaufort Scale describes everything the wind can do in 110 words—somewhere there may be a better example of "So many *things*, almost in an equal number of *words*." But if there is, I've never found it.

IF THE *MANUAL OF SCIENTIFIC ENQUIRY* is a kind of grown-up *Boy Scout Handbook*, Beaufort's life is its object lesson. And the Beaufort Scale, at last, is the derivative, the essence, the shorthand for that entire way of life. Just the way clear thinking and clear writing have a one-to-one ratio—you can't have one without the other—the Beaufort Scale has that kind of relationship to an observant, attentive life. If you're thinking about things in a Beaufort Scale way, you can't fail to pay attention. The words in the scale are like the life Beaufort—and Dalrymple, and Smeaton, and all the others—lived: clean, crisp, everything useful, everything used. I eventually found myself comparing my voyage into the life of Beaufort and the history of the wind scale to the voyages Beaufort made himself, especially his hydrographic journey to Karamania, where he found antiquities and points of interest so fascinating he just had to tell everyone about them. Beaufort went to points unexplored since classical times; I found myself up to my elbows in papers and stories of men far too little remembered. Beaufort used a copy of Strabo's geography as his guide; I used as my guide Beaufort's own book, as well as the *Manual of Scientific Enquiry* and, above all, the Beaufort Scale.

The point of this enterprise? That's easy—in the preface of *Karamania,* his book about his own journey, Beaufort states the case far better than I can. He notes that his work is, like all work, based on the efforts of those who have gone before; he points out that though he was unable to complete to his own satisfaction the hydrographical task he had set for himself, he nonetheless couldn't fail to investigate things he found along the way, things "too numerous and too interesting not to have found some admission among those remarks."

And finally he says that his results could never be complete or authoritative, but since they covered something so little known he felt obligated to share them with the public—"not indeed with the vain expectation of satisfying curiosity," he said, "but rather in the hope of exciting further inquiry . . . ; and if they throw but little light on antient history, or add still less to modern science, they may perhaps rouse others to visit this, hitherto, neglected country, whose leisure and whose talents are better adapted to those pursuits." And as I have found time and again with the Beaufort Scale, these words apply to my own life, and to the work at hand.

That's a kind of open-hearted intellectual decency to which we ought to aspire every single day: That's the voice of a man who believes that the answers are out there if we just go looking for them. That's the voice of a man who believes that nature, rightly questioned, never lies, and that it is incumbent on us all to get up early in the morning to start formulating the best questions we can, and write down what we learn in a way that everyone can understand.

I don't know if you would call that behavior honor, but you might well call it spirit—and it's a Beaufort kind of spirit. You could probably use it as a characteristic to gauge value—as a sort of metric, something to look for in what you read, the people you meet; to determine their value, and to compare what you read and saw with other things you have read and seen. It could be useful, something you could memorize, to help you remind yourself to seek those characteristics. Maybe someone will even make a scale about that.

I wonder what they could call it.

APPENDIX A

Beaufort Scale Family Album

1. This list of terms from Defoe's *The Storm*, of 1704, shows that by the early 1700s descriptions of the wind were finding some organization.

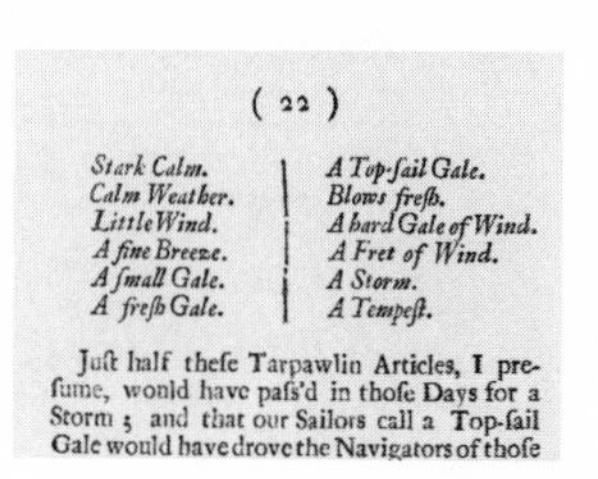

(22)

Stark Calm.	*A Top-ſail Gale.*
Calm Weather.	*Blows freſh.*
Little Wind.	*A hard Gale of Wind.*
A fine Breeze.	*A Fret of Wind.*
A ſmall Gale.	*A Storm.*
A freſh Gale.	*A Tempeſt.*

Juſt half theſe Tarpawlin Articles, I preſume, would have paſs'd in thoſe Days for a Storm ; and that our Sailors call a Top-ſail Gale would have drove the Navigators of thoſe

Source: The Storm by Daniel DeFoe, p. 22. Special Collections Library, University of Michigan.

2. In 1759, engineer John Smeaton added wind speed to this wind scale in his *Experimental Enquiry Concerning the Natural Powers of Wind and Water,* though the descriptive terms weren't yet completely standardized.

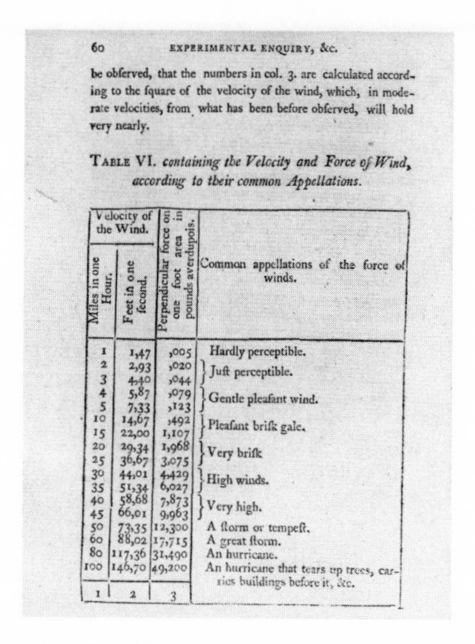

60 EXPERIMENTAL ENQUIRY, &c.

be obſerved, that the numbers in col. 3. are calculated according to the ſquare of the velocity of the wind, which, in moderate velocities, from what has been before obſerved, will hold very nearly.

TABLE VI. *containing the Velocity and Force of Wind, according to their common Appellations.*

Velocity of the Wind. Miles in one Hour.	Feet in one ſecond.	Perpendicular force on one foot area in pounds averdupois.	Common appellations of the force of winds.
1	1,47	,005	Hardly perceptible.
2	2,93	,020	Juſt perceptible.
3	4,40	,044	
4	5,87	,079	Gentle pleaſant wind.
5	7,33	,123	
10	14,67	,492	Pleaſant briſk gale.
15	22,00	1,107	
20	29,34	1,968	Very briſk
25	36,67	3,075	
30	44,01	4,429	High winds.
35	51,34	6,027	
40	58,68	7,873	Very high.
45	66,01	9,963	
50	73,35	12,300	A ſtorm or tempeſt.
60	88,02	17,715	A great ſtorm.
80	117,36	31,490	An hurricane.
100	146,70	49,200	An hurricane that tears up trees, carries buildings before it, &c.
1	2	3	

Source: From *Experimental Enquiry Concerning the Natural Powers of Wind and Water . . .* by John Smeaton. Special Collections Library, University of Michigan.

3. Smeaton eventually applied his terms and speeds to an observable, measurable object: a windmill. Around 1790, hydrographer Alexander Dalrymple connected Smeaton's windmill scale with sea terms.

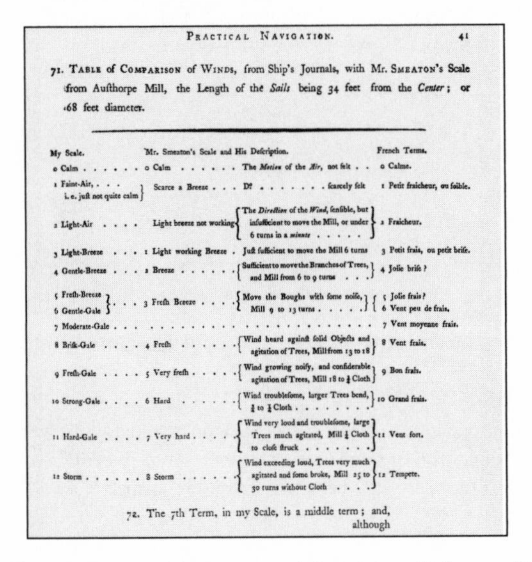

PRACTICAL NAVIGATION. 41

71. TABLE of COMPARISON of WINDS, from Ship's Journals, with Mr. SMEATON's Scale from Austhorpe Mill, the Length of the *Sails* being 34 feet from the *Center*; or 68 feet diameter.

My Scale.	Mr. Smeaton's Scale and His Description.		French Terms.
0 Calm	0 Calm	The *Motion* of the *Air*, not felt	0 Calme.
1 Faint-Air, i. e. just not quite calm	Scarce a Breeze	D° scarcely felt	1 Petit fraicheur, ou foible.
2 Light-Air	Light breeze not working	The *Direction* of the *Wind*, sensible, but insufficient to move the Mill, or under 6 turns in a *minute*	2 Fraicheur.
3 Light-Breeze	1 Light working Breeze	Just sufficient to move the Mill 6 turns	3 Petit frais, ou petit brise.
4 Gentle-Breeze	2 Breeze	Sufficient to move the Branches of Trees, and Mill from 6 to 9 turns	4 Jolie brise?
5 Fresh-Breeze 6 Gentle-Gale	3 Fresh Breeze	Move the Boughs with some noise, Mill 9 to 13 turns	5 Jolie frais? 6 Vent peu de frais.
7 Moderate-Gale			7 Vent moyenne frais.
8 Brisk-Gale	4 Fresh	Wind heard against solid Objects and agitation of Trees, Mill from 13 to 18	8 Vent frais.
9 Fresh-Gale	5 Very fresh	Wind growing noisy, and considerable agitation of Trees, Mill 18 to ¾ Cloth	9 Bon frais.
10 Strong-Gale	6 Hard	Wind troublesome, larger Trees bend, ¾ to ½ Cloth	10 Grand frais.
11 Hard-Gale	7 Very hard	Wind very loud and troublesome, large Trees much agitated, Mill ½ Cloth to close struck	11 Vent fort.
12 Storm	8 Storm	Wind exceeding loud, Trees very much agitated and some broke, Mill 25 to 30 turns without Cloth	12 Tempete.

72. The 7th Term, in my Scale, is a middle term; and, although

Source: Used by permission of the Trustees of the National Library of Scotland.

4. In 1806, cribbing from an early version of Dalrymple's list, Francis Beaufort copied these terms into his journal—and used the terms to keep a journal of the wind until the end of his life.

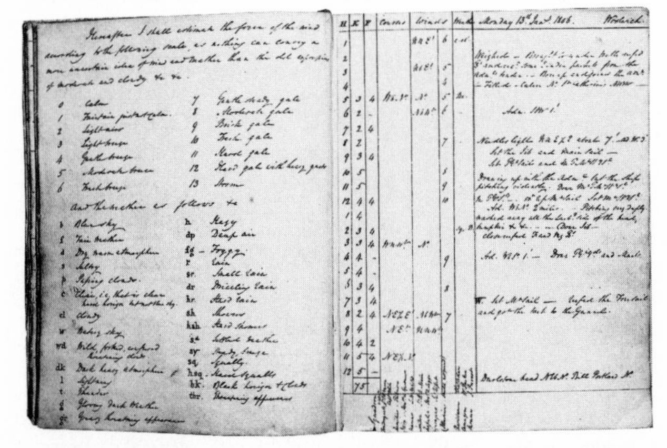

0 Calm
2 Light air
3 Light breeze
4 Gentle breeze
5 Moderate breeze
6 Fresh breeze
7 Gentle steady gale
8 Moderate gale
9 Brisk gale
10 Fresh gale
11 Hard gale
12 Hard gale with heavy gusts
13 Storm

Monday 13th Jan. 1806.

Source: From the journals of Francis Beaufort, Met Office, National Meteorological Archive. Used by permission.

5. Seeking to make the scale as useful as possible, Beaufort quickly followed—knowingly or unknowingly—Smeaton's trick of applying the wind names to an observable object, in this case, the very ship on which he sailed. This updated scale appeared at the front of a journal in 1807 or 1808.

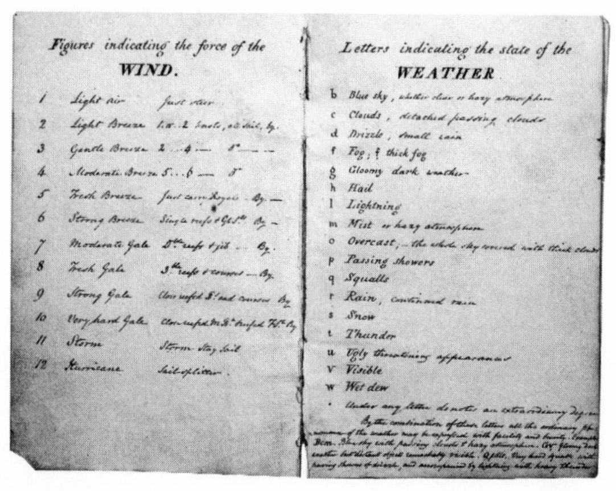

Figures indicating the force of the
WIND.

1 Light air
2 Light Breeze
3 Gentle Breeze
4 Moderate Breeze
5 Fresh Breeze
6 Strong Breeze
7 Moderate Gale
8 Fresh Gale
9 Strong Gale
10 Very hard Gale
11 Storm — Storm stay sail
12 Hurricane

Letters indicating the state of the
WEATHER

b Blue sky, whether clear or hazy atmosphere
c Clouds, detached passing clouds
d Drizzle, small rain
f Fog; thick fog
g Gloomy dark weather
h Hail
l Lightning
m Mist or hazy atmosphere
o Overcast, — the whole sky covered with thick clouds
p Passing showers
q Squalls
r Rain, continued rain
s Snow
t Thunder
u Ugly threatening appearance
v Visible
w Wet dew
· Under any letter denotes an extraordinary degree

Source: From the journals of Francis Beaufort, Met Office, National Meteorological Archive. Used by permission.

6. Beaufort codified his state-of-sail scale to make it widely useful. This version, with Beaufort's initials typeset in the lower right, was apparently used in the 1830s to lobby for the scale's widespread adoption.

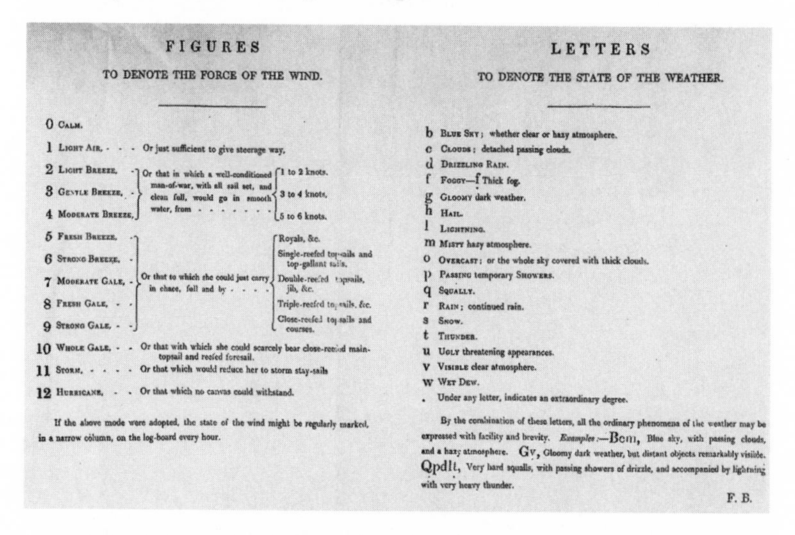

FIGURES

TO DENOTE THE FORCE OF THE WIND.

0 Calm.
1 Light Air, - - - Or just sufficient to give steerage way.
2 Light Breeze, - } Or that in which a well-conditioned man-of-war, with all sail set, and clean full, would go in smooth water, from - - - - - - - { 1 to 2 knots.
3 Gentle Breeze, - } { 3 to 4 knots.
4 Moderate Breeze, } { 5 to 6 knots.
5 Fresh Breeze, - } Or that to which she could just carry in chace, full and by - - - - { Royals, &c.
6 Strong Breeze, - } { Single-reefed top-sails and top-gallant sails.
7 Moderate Gale, - } { Double-reefed topsails, jib, &c.
8 Fresh Gale, - - } { Triple-reefed topsails, &c.
9 Strong Gale, - - } { Close-reefed topsails and courses.
10 Whole Gale, - - Or that with which she could scarcely bear close-reefed main-topsail and reefed foresail.
11 Storm, - - - - Or that which would reduce her to storm stay-sails
12 Hurricane, - - Or that which no canvas could withstand.

If the above mode were adopted, the state of the wind might be regularly marked, in a narrow column, on the log-board every hour.

LETTERS

TO DENOTE THE STATE OF THE WEATHER.

b Blue Sky; whether clear or hazy atmosphere.
c Clouds; detached passing clouds.
d Drizzling Rain.
f Foggy—f̣ Thick fog.
g Gloomy dark weather.
h Hail.
l Lightning.
m Misty hazy atmosphere.
o Overcast; or the whole sky covered with thick clouds.
p Passing temporary Showers.
q Squally.
r Rain; continued rain.
s Snow.
t Thunder.
u Ugly threatening appearances.
v Visible clear atmosphere.
w Wet Dew.
. Under any letter, indicates an extraordinary degree.

By the combination of these letters, all the ordinary phenomena of the weather may be expressed with facility and brevity. *Examples:*—Bcm, Blue sky, with passing clouds, and a hazy atmosphere. Gṿ, Gloomy dark weather, but distant objects remarkably visible. Qp̣dlṭ, Very hard squalls, with passing showers of drizzle, and accompanied by lightning with very heavy thunder.

F. B.

Source: Met Office, National Meteorological Archive. Used by permission.

7. By 1870, the Smithsonian Institution was sending its observers this wind scale.

Wind Scale, Smithsonian Institution, 1870

0. A perfect calm. The simple letter of the wind, for instance N. (north), indicating its direction without any number, means a slight movement of the air hardly to be called a wind, and only just sufficient to allow an estimate of its direction.
1. A light breeze which moves the foliage, and sometimes fans the face.
2. A wind which moves the branches of the trees, somewhat retards walking, and causes more or less of a slight rustling sound in the open air.
3. A wind which causes strong boughs and entire trees to rock, makes walking against it difficult; which causes a stronger rustling sound to be heard, and which often blows in gusts, and carries light bodies up into the air.
4. A storm-wind, during which the trees are in constant motion; branches and boughs covered with foliage are broken off, and in a violent storm sometimes even entire trees are broken, or uprooted; leaves, dust & c., are continually borne up and carried far away; during which time there is an uninterrupted loud rustling sound, whith strong gusts; walking windward is extremely difficult, and now and then chimneys, fences & c., are thrown down, windows broken in, & c.

Source: Directions for Meteorological Observations and the Registry of Periodical Phenomena (Washington, DC: Smithsonian, 1870).

8. Around 1900, seismologists followed meteorologists and developed a descriptive scale to categorize earthquakes.

The Modified Mercalli Earthquake Intensity Scale

I. Not felt. Marginal and long-period effects of large earthquakes.
II. Felt by persons at rest, on upper floors, or favorably placed.
III. Felt indoors. Hanging objects swing. Vibration like passing of light trucks. Duration estimated. May not be recognized as an earthquake.
IV. Hanging objects swing. Vibration like passing of heavy trucks; or sensation of a jolt like a heavy ball striking the walls. Standing cars rock. Windows, dishes, doors rattle. Glasses clink. Crockery clashes. In the upper range of IV, wooden walls and frame creak.
V. Felt outdoors; direction estimated. Sleepers awakened. Liquids disturbed, some spilled. Small unstable objects displaced or upset. Doors swing, close, open. Shutters, pictures move. Pendulum clocks stop, start, change rate.
VI. Felt by all. Many frightened and run outdoors. Persons walk unsteadily. Windows, dishes, glassware broken. Knickknacks, books, etc., off shelves. Pictures off walls. Furniture moved or overturned. Weak plaster and masonry D cracked. Small bells ring (church, school). Trees, bushes shaken visibly, or heard to rustle.
VII. Difficult to stand. Noticed by drivers. Hanging objects quiver. Furniture broken. Damage to masonry D, including cracks. Weak chimneys broken at roof line. Fall of plaster, loose bricks, stones, tiles, cornices, also unbraced parapets

and architectural ornaments. Some cracks in masonry C. Waves on ponds, water turbid with mud. Small slides and caving in along sand or gravel banks. Large bells ring. Concrete irrigation ditches damaged.

VIII. Steering of cars affected. Damage to masonry C; partial collapse. Some damage to masonry B; none to masonry A. Fall of stucco and some masonry walls. Twisting, fall of chimneys, factory stacks, monuments, towers, elevated tanks. Frame houses moved on foundations if not bolted down; loose panel walls thrown out. Decayed piling broken off. Branches broken from trees. Changes in flow or temperature of springs and wells. Cracks in wet ground and on steep slopes.

IX. General panic. Masonry D destroyed; masonry C heavily damaged, sometimes with complete collapse; masonry B seriously damaged. General damage to foundations. Frame structures, if not bolted, shifted off foundations. Frames racked. Serious damage to reservoirs. Underground pipes broken. Conspicuous cracks in ground. In alluviated areas, sand and mud ejected, earthquake fountains, sand craters.

X. Most masonry and frame structures destroyed with their foundations. Some well-built wooden structures and bridges destroyed. Serious damage to dams, dikes, embankments. Large landslides. Water thrown on banks of canals, rivers, lakes, etc. Sand and mud shifted horizontally on beaches and flat land. Rails bent slightly.

XI. Rails bent greatly. Underground pipelines completely out of service.

XII. Damage nearly total. Large rock masses displaced. Lines of sight and level distorted. Objects thrown into the air.

Source: U.S. Geological Survey (http://hvo.wr.usgs.gov/earthquakes/felt/mercalli.html).

9. In 1906, exactly one hundred years after Beaufort copied the Dalrymple scale into his journal, the Met Office of England created the modern Beaufort Scale, replete with poetics, in this table from Official Publication 180.

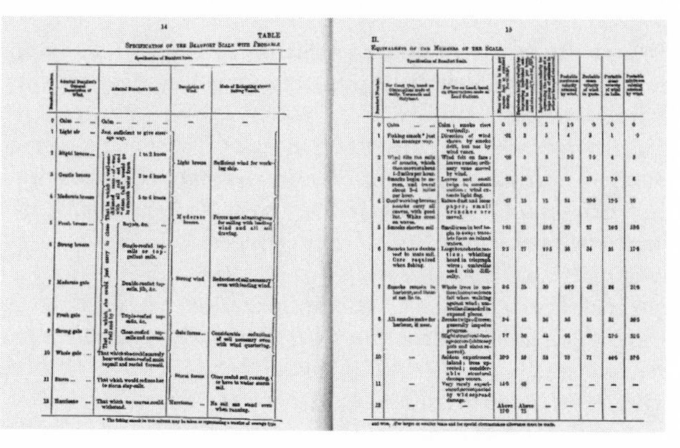

Source: Met Office, National Meteorological Archive. Used by permission.

10. In 1927, in response to the decreasing utility of a scale based on sail use, the German captain Petersen recast the scale based on the state of the sea.

Beaufort Scale: Petersen State-of-Sea Version

Force 0: Calm; mean speed of 0 knots
Sea like a mirror.
Force 1: Light air; mean speed 2 knots
Ripples with the appearance of scales are formed, but without foam crests.
Force 2: Light breeze; 5 knots
Small wavelets, still short but more pronounced. Crests have a glassy appearance and do not break.
Force 3: Gentle breeze; 9 knots
Large wavelets. Crests begin to break. Foram of glassy appearance. Perhaps scattered white horses.
Force 4: Moderate breeze; 13 knots
Large waves begin to form; the white foam crests are more extensive everywhere. Probably some spray.
Force 5: Fresh breeze; 19 knots
Moderate waves, taking a more pronounced long form; many white horses are formed. Chance of some spray.
Force 6: Strong breeze: 24 knots
Large waves begin to form; the white foam crests are more extensive everywhere. Probably some spray.
Force 7: Near gale; 30 knots
Sea heaps up and white foam from breaking waves begins to be blown in streaks along the direction of the wind.
Force 8: Gale; 37 knots.
Moderately high waves of greater length; edges of crests begin to break into spindrift. The foam is blown in well-marked streaks along the direction of the wind.
Force 9: Strong gale; 44 knots.
High waves. Dense streaks of foam along the direction of the wind. Crests of waves begin to topple, tumble and roll over. Spray may affect visibility.
Force 10: Storm; 52 knots.
Very high waves with long overhanging crests. The resulting foam, in great patches, is blown in dense white streaks along the direction of the wind. On the whole, the surface of the sea takes a white appearance. The "tumbling" of the sea becomes heavy and shock-like. Visibility extended.
Force 11: Violent storm; 60 knots
Exceptionally high waves (small and medium-sized ships might be lost to view for a time behind the waves). The sea is completely covered with long white patches of foam lying along the direction of the wind. Everywhere the edges of the wave crests are blown into froth. Visibility affected.
Force 12: Hurricane: 64 knots and over.

The air is filled with foam and spray. Sea completely white with driving spray; visibility seriously affected.

Source: The Met Office (www.metffice.gov.uk/education/curriculum/leaflets/beaufort.html).

11. The Merriam-Webster Collegiate dictionaries began including the Beaufort Scale in the 1940s. It remained virtually unchanged there until 1993.

beat ● because

ated striking **e**
; *esp* : to flatten
rumbeat **3** : to
ME, DEFEAT; *also*
despite ⟨~ the
XHAUST (2) : to
, SWINDLE **5 a**
to report a news
c : CIRCUMVENT
tempo⟩ ~ *vi* **1**
glare or strike
ting activity **d**
2) : TICK **b** : to
d blows **b** : to
rouse game; *also*
o progress with
at around the
course — beat it
at one's brains
t by thinking —
all possible areas
connected with

ies; *also* : PULSA-
ting **c** : a driv-
n or balance of a
produced by the
having different
g or foot against
mic stress in po-
tresses **b** : the
al performer **c**
: driving force in
traversed round
⟨I've never seen
head of competi-
o windward **b**
ss \-ləs\ *adj*
being in a state of
n or morale **2**

l shape ⟨~ gold⟩
LIAR ⟨a ~ path⟩

GGBEATER **b** : a
DRUMSTICK 1 **2**
e game **3** : an

happy, fr. *beatus*
good — more at
de **2** : having a

BEAUFORT SCALE

BEAUFORT NUMBER	NAME	MILES PER HOUR	DESCRIPTION
0	calm	less than 1	calm; smoke rises vertically
1	light air	1–3	direction of wind shown by smoke but not by wind vanes
2	light breeze	4–7	wind felt on face; leaves rustle; ordinary vane moved by wind
3	gentle breeze	8–12	leaves and small twigs in constant motion; wind extends light flag
4	moderate breeze	13–18	raises dust and loose paper; small branches are moved
5	fresh breeze	19–24	small trees in leaf begin to sway; crested wavelets form on inland waters
6	strong breeze	25–31	large branches in motion; telegraph wires whistle; umbrellas used with difficulty
7	moderate gale (*or* near gale)	32–38	whole trees in motion; inconvenience in walking against wind
8	fresh gale (*or* gale)	39–46	breaks twigs off trees; generally impedes progress
9	strong gale	47–54	slight structural damage occurs; chimney pots and slates removed
10	whole gale (*or* storm)	55–63	trees uprooted; considerable structural damage occurs
11	storm (*or* violent storm)	64–72	very rarely experienced; accompanied by widespread damage
12	hurricane*	73–136	devastation occurs

*The U.S. uses 74 statute mph as the speed criterion for hurricane.

syn BEAUTIFUL, LOVELY, HANDSOME, PRETTY, COMELY, FAIR *shared meaning element* : pleasing to the mind, spirit, or senses ***ant*** ugly

Source: By permission, from Webster's Ninth *New Collegiate Dictionary* © 1983 by Merriam Webster, Incorporated (www.Merriam-Webster.com).

12. In 1971, University of Chicago professor Theodore Fujita created for tornadoes what the Beaufort Scale was for wind.

Fujita Tornado Damage Scale

F0; winds below 73 mph
Light damage. Some damage to chimneys; branches broken off trees; shallow-rooted trees pushed over; sign boards damaged.
F1; winds 73–112 mph
Moderate damage. Peels surface off roofs; mobile homes pushed off foundations or overturned; moving autos blown off roads.
F2; winds 113–157 mph
Considerable damage. Roofs torn off frame houses; mobile homes demolished; boxcars overturned; large trees snapped or uprooted; light-object missiles generated; cars lifted off ground.
F3; winds 158–206
Severe damage. Roofs and some walls torn off well-constructed houses; trains overturned; most trees in forest uprooted; heavy cars lifted off the ground and thrown.
F4; winds 207–260 mph
Devastating damage. Well-constructed houses leveled; structures with weak foundations blown away some distance; cars thrown and large missiles generated.
F5; winds 261–318 mph
Incredible damage. Strong frame houses leveled off foundations and swept away; automobile-sized missiles fly through the air in excess of 100 meters (109 yds); trees debarked; incredible phenomena will occur.

Source: National Oceanographic and Atmospheric Administration.

13. In 1974, Herbert Saffir, a Florida engineer, and Robert Simpson, of the National Hurricane Center in Miami, prepared a Hurricane Disaster Potential Scale of five levels, now known as the Saffir-Simpson Scale.

The Saffir-Simpson Hurricane Scale

Category One Hurricane

Winds 74–95 mph (64–82 kt or 119–153 km/hr). Storm surge generally 4–5 ft above normal. No real damage to building structures. Damage primarily to unanchored mobile homes, shrubbery, and trees. Some damage to poorly constructed signs. Also, some coastal road flooding and minor pier damage.

Category Two Hurricane

Winds 96–110 mph (83–95 kt or 154–177 km/hr). Storm surge generally 6–8 feet above normal. Some roofing material, door, and window damage of buildings. Considerable damage to shrubbery and trees with some trees blown down. Considerable damage to mobile homes, poorly constructed signs, and piers.

Coastal and low-lying escape routes flood 2–4 hours before arrival of the hurricane center. Small craft in unprotected anchorages break moorings.

Category Three Hurricane

Winds 111–130 mph (96–113 kt or 178–209 km/hr). Storm surge generally 9–12 ft above normal. Some structural damage to small residences and utility buildings with a minor amount of curtainwall failures. Damage to shrubbery and trees with foliage blown off trees and large trees blown down. Mobile homes and poorly constructed signs are destroyed. Low-lying escape routes are cut by rising water 3–5 hours before arrival of the center of the hurricane. Flooding near the coast destroys smaller structures with larger structures damaged by battering from floating debris. Terrain continuously lower than 5 ft above mean sea level may be flooded inland 8 miles (13 km) or more. Evacuation of low-lying residences with several blocks of the shoreline may be required.

Category Four Hurricane

Winds 131–155 mph (114–135 kt or 210–249 km/hr). Storm surge generally 13–18 ft above normal. More extensive curtainwall failures with some complete roof structure failures on small residences. Shrubs, trees, and all signs are blown down. Complete destruction of mobile homes. Extensive damage to doors and windows. Low-lying escape routes may be cut by rising water 3–5 hours before arrival of the center of the hurricane. Major damage to lower floors of structures near the shore. Terrain lower than 10 ft above sea level may be flooded requiring massive evacuation of residential areas as far inland as 6 miles (10 km).

Category Five Hurricane

Winds greater than 155 mph (135 kt or 249 km/hr). Storm surge generally greater than 18 ft above normal. Complete roof failure on many residences and industrial buildings. Some complete building failures with small utility buildings blown over or away. All shrubs, trees, and signs blown down. Complete destruction of mobile homes. Severe and extensive window and door damage. Low-lying escape routes are cut by rising water 3–5 hours before arrival of the center of the hurricane. Major damage to lower floors of all structures located less than 15 ft above sea level and within 500 yards of the shoreline. Massive evacuation of residential areas on low ground within 5–10 miles (8–16 km) of the shoreline may be required.

Source: National Oceanographic and Atmospheric Administration.

14. The original source of the changes to the Beaufort Scale introduced in *Merriam-Webster's Collegiate Dictionary*, Tenth Edition, remains somewhat mysterious.

ESTIAL 1 2
; *n*
avy loads or
[ME *beten,*
1 : to strike
en used with
orcefully and
usly **e** : to
as if in quest
n *up* **g** : to
a drum⟩ **2**
er, paste, or
d (1) : to
peated strik-
esp : to flat-
beat **3** : to
T; *also* : SUR-
he odds⟩ **c**
leave dispir-
(1) : to act
n in advance
system⟩ **d**
e against (a
vi **1 a** : to
with oppres-
at a drum **2**
ng struck **3**
c : to strike
r or as if for
windward by
about the
: to the point
at it 1 : to
ins out : to
— **beat the**
— **beat the**
plicize vigor-
es connected
physically or
es; *also* : PUL-
c : a driving
ce of a time-
ie ~⟩ **b** : a
a : a metri-
effect of these
musical per-
ristic driving
t excels ⟨I've
head of com-
ard **b** : one
ons of ampli-
electric cur-

(1920) : DANDY 1
beau•coup \'bō-(ˌ)kü\ *adj* [F] (1918) *slang* : great in quantity or amount : MANY, MUCH ⟨spent ~ dollars⟩
Beau•fort scale \'bō-fərt-\ *n* [Sir Francis *Beaufort*] (1858) : a scale in which the force of the wind is indicated by numbers from 0 to 12

BEAUFORT SCALE

BEAUFORT NUMBER	NAME	WIND SPEED MPH	WIND SPEED KPH	DESCRIPTION
0	calm	<1	<1	calm; smoke rises vertically
1	light air	1-3	1-5	direction of wind shown by smoke but not by wind vanes
2	light breeze	4-7	6-11	wind felt on face; leaves rustle; wind vane moves
3	gentle breeze	8-12	12-19	leaves and small twigs in constant motion; wind extends light flag
4	moderate breeze	13-18	20-28	wind raises dust and loose paper; small branches move
5	fresh breeze	19-24	29-38	small-leaved trees begin to sway; crested wavelets form on inland waters
6	strong breeze	25-31	39-49	large branches move; overhead wires whistle umbrellas difficult to control
7	moderate gale *or* near gale	32-38	50-61	whole trees sway; walking against wind is difficult
8	fresh gale *or* gale	39-46	62-74	twigs break off trees; moving cars veer
9	strong gale	47-54	75-88	slight structural damage occurs; shingles may blow away
10	whole gale *or* storm	55-63	89-102	trees uprooted; considerable structural damage occurs
11	storm *or* violent storm	64-72	103-117	widespread damage occurs
12	hurricane*	>72	>117	widespread damage occurs

*The U.S. uses 74 statute mph as the speed criterion for a hurricane.

Source: By permission, from *Merriam-Webster's Collegiate Dictionary*, Tenth Edition

APPENDIX B

Explanation of the Plate Describing the Rigging, &c., of a First-Rate Man of War

348

EXPLANATION of the PLATE describing the RIGGING, &c., of a FIRST-RATE MAN OF WAR.

1 BOWSPRIT
2 Gammoning
3 Cap
4 Bobstay
5 Manrope
6 Spritsail yard
7 Lifts
8 Standing lifts.
9 Horses
10 Parrel
11 Braces and Pendants
12 Sheets and pendants
13 Cluelines
14 Buntlines
15 Jib-boom
16 Traveller
17 Horse
18 Stay
19 Haliards
20 Guy
21 Jack-staff
22 Truck
23 Jack-flag

Fore, main, and mizen-mast, rigged alike, as on the top-mast and top-gallant mast, and all the yards, except the cross-jack yard, which has no sail; therefore the description of one serves for the other, except where otherwise expressed.

24 Foremast
25 Woolding
26 Fish
27 Top
28 Cap
29 Runner and tackle
30 Shrouds
31 Laniards
32 Ratlines
33 Stay and laniard
34 Spring stay and ditto
35 Snakeline
36 Crowfoot
37 Fore yard
38 Geers
39 Lifts
40 Braces and pendants
41 Cluelines
42 Buntlines
43 Horses and stirrups
44 Leechlines
45 Yard tackles
46 Bowlines and bridles
47 Tacks
48 Sheets
49 Truss parrel
50 Pudding
51 Dolphin
52 Top-rope
53 Topmast
54 Crosstrees
55 Cap
56 Runner
57 Shrouds and laniards
58 Stays
59 Backstays
60 Staysail haliards
61 Topsail yard
62 Tye and haliard
63 Lifts
64 Braces and pendants
65 Horses
66 Parrel
67 Flemish horse
68 Buntlines
69 Cluelines
70 Bowlines and bridles
71 Reef tackles and pendants
72 Jewel blocks
73 Sheets
74 Top-gallant mast
75 Shrouds
76 Stay
77 Backstay
78 Top-gallant yard
79 Haliard
80 Lifts
81 Horse
82 Parrel
83 Clueline
84 Bowline
85 Sheet
86 Royal mast
87 Stay
88 Backstay
89 Truck
90 Admiralty flag
91 Middle-stay-sail stay
92 Haliards
93 Top-gal. stay-sail haliards
94 Mizen gaff.
95 Derrick and span
96 Peek brails
97 Spanker haliards
98 Vangs
99 Cross-jack yard
100 Spanker-boom
101 Topping-lift
102 Poop lantern
103 Stern ladder
104 Rudder chains
105 Standard flag
106 Union flag
107 Ensign staff
108 Ensign flag
109 Puttock-shrouds
110 Cable

HULL.

A Head or stem
B Forecastle
C Waist
D Quarter-deck
E Poop
F Stern or abaft

Source: From *The Practical Navigator,* p. 348. Graduate Library, University of Michigan.

A Note on the Sources

Footnotes would have rendered this book unreadable and possibly unwriteable. I wandered among such scattered sources, and was at such a loss among them for such a long time, that it seems almost a little sad to organize them now. Or rather, as I do so, it makes me want to start writing the book over again now that I can finally see something like the lay of the land. I guess it's usually that way.

The oldest source I consulted for this book was Aristotle's *Meteorologica,* the Loeb Classical Library edition (Cambridge, MA, 1987), and the most recent is the Shipping Forecast, which was most recently updated sometime in the last six hours or so. You can hear it four times a day on BBC Radio 4. As for the Aristotle, if you can't read it in the British Museum Reading Room, as I did (if only to share a fold-out reader's desk with Marx or Twain or Yeats), you can find it in the rotunda of the Library of Congress or, if you have to, at just about any bookstore.

In most cases I do not list dictionaries or encyclopedias because I used every edition of every dictionary I could find, and the most current editions of most encyclopedias, all of which are available at almost any library, though the 1911 Eleventh Edition of the *Encyclopaedia Britannica* is worth whatever you have to do to find it.

Regarding dictionaries, readers are of course warned that "Webster's" means nothing; the word has passed into the lexicon. Only the Merriam-Webster dictionaries of Springfield, Massachusetts, are the legitimate intellectual descendants of Noah Webster. The Webster's of 1828, the *Webster's International* of 1890, the *New International* of 1909, the Second Edition of 1934, and the Third Edition of 1961—plus of course the *Collegiate*

Dictionaries that are their authorized abridgements—are the dictionaries I most thoroughly discussed in this book.

The bibliographies of the two biographies of Francis Beaufort (by Friendly and Courtney) proved so helpful to me that they became in themselves invaluable sources. So, in the hope that anyone motivated to try to go where I did not go, figure out what I could not, or correct mistakes I have made will find at least some direction here, I list the sources I consulted in the different general areas of this book. Readers or researchers seeking more detailed information they suspect I may possess are welcome to contact me through the publisher. Books so cool that you absolutely have to read them are printed in **boldface type.**

Regarding Admiral Sir Francis Beaufort

Beaufort, Francis. *Karamania.* London: Hunter, 1817.

Collins, K. St. B. "Admiral Sir Francis Beaufort." *Journal of the Institute of Navigation,* 1958.

Courtney, Nicholas. "Beaufort's Hunt." *Cornucopia* 5, no. 27 (2002).

———. *Gale Force* 10. London: Review, 2002.

Ellison, C. C. *The Hopeful Traveler.* Dublin, 1987.

Friendly, Alfred. *Beaufort of the Admiralty.* New York: Random House, 1977.

Ireland, J. de Courcy. "Frances Beaufort, Irish Hydrographer." *Technology Ireland* (July/August 1974).

Martineau, Harriett. "Notice of Rear-Admiral Sir Francis Beaufort, K.C.B." Reprinted from *Daily News,* London, January 15, 1858.

Regarding ships, the history of navigation, and chartmaking

American Sail Training Association. *Sail Tall Ships!* 13th ed. Newport, RI, 2001.

Azcel, Amir. *The Riddle of the Compass.* New York: Harvest, 2001.

Bauer, Bruce. *The Sextant Handbook.* 2d. ed. Camden, ME: Ragged Mountain Press, 1995.

Bellin, Jacques Nicolas. *Le petit atlas maritime, Recueil de cartes et plans des quatre parties du monde.* Paris: 1764.

Brown, Lloyd. *The Story of Maps.* Boston: Little, Brown, 1949.

Cain, M. T. "The Maps of the Society for the Diffusion of Useful Knowledge: A Publishing History." *Imago Mundi* 46 (1994).

Chapuis, Oliver. *A la mer au ciel: Beautemps-Beaupre et la renaissance de l'hydrographie moderne (1700–1850). L'Emergence de la precision en navigation et dans la cartographie marine.* Paris: Musee Nationale de la Marine, 1999.

Collinder, Per. *A History of Marine Navigation.* London: Batsford, 1954.

Falconer, William. *The Shipwreck.* London: Walker, 1811.

Frampton, R. M., and P. A. Uttridge. *Meteorology for Seafarers.* Glasgow: Brown, Son, & Ferguson, 1988.

Hannay, David. *The Navy and Sea Power.* London: Thornton Butterworth, 1913.

Howse, Derek, ed. *Five Hundred Years of Nautical Science,* 1400–1900. London: National Maritime Museum, 1981.

Jeans, Peter D. *Ship to Shore: A Dictionary of Everyday Words and Phrases Derived from the Sea.* Oxford: ABC-Clio, 1998.

Kemp, Peter, ed. *The Oxford Companion to Ships and the Sea.* Oxford: Oxford University Press, 1976.

Marine Observer's Handbook. 11th ed. London: Met Office, 1995.

Mariner's Handbook. 6th ed. London: Hydrographer of the Navy, 1989.

Moore, John Hamilton. *The Practical Navigator; Being a Complete Epitome of Navigation.* 20th ed. London: Cadell et al., 1828.

"Narrative Concerning the Success of Pendulum-Watches at Sea for the Longitudes." *Philosophical Transactions.* vol. 1, 1665–1666, 13–15.

O'Brian, Patrick. *Master and Commander.* New York: Norton, 1990.

Remarkable Shipwrecks: A Collection of Interesting Accounts of Naval Disasters with Many Particulars . . . Together with an Account of the Deliverance of Survivors Selected from Authentic Sources. Hartford, CT.: Andrews & Starr, 1813.

Riesenberg, Felix. *Standard Seamanship for the Merchant Service.* New York: Van Nostrand, 1936.

Ritchie, G. S. *The Admiralty Chart.* New York: American Elsevier, 1967.

———. "Great Britain's Contribution to Hydrography During the Nineteenth Century." *Institute of Navigation Journal* 20, no. 1 (January 1967).

Robinson, A. H. W. *Marine Cartography in Britain.* Leicester University Press, 1962.

Smythe, William Henry. *The Sailor's Word Book.* London: Blackie & Son, 1867.

Sobel, Dava. *Longitude.* New York: Walker, 1995.

Southworth, Michael and Susan. *Maps.* Boston: Little, Brown, 1982.

Taylor, E. G. R., and M. W. Richey. *The Geometrical Seaman: A Book of Early Nautical Instruments.* London: Hollis & Carter, 1962.

Thompson, Silvanus P. *The Rose of the Winds.* London: Oxford University Press, 1913.

Wilford, John Noble. *The Mapmakers.* New York: Knopf, 1981.

Regarding the wind and how it is understood

Andersson, Tage. "Admiral Johan Henrik Kreuger and his Anemometer." Monograph, Swedish Meteorological Society, 2003.

Blumenstock, David. *The Ocean of Air.* New Brunswick, NJ: Rutgers, 1959.

Bohun, Ralph. *A Discourse Concerning the Origine and Properties of Wind. With an Historicall Account of Hurricanes, and Other Tempestuous Winds.* Oxford: 1671.

Brice, Alex. "A Letter to the President of the Royal Society, Containing a New Manner of Measuring the Velocity of Wind, and an Experiment to Ascertain to What Quantity of Water a Fall of Snow is Equal." *Philosophical Transactions* 56 (1766), 224–29.

DeBlieu, Jan. *The Wind.* Boston, 1998.

Defoe, Daniel. *The Storm, or a Collection of the most remarkable casualties and disasters which happen'd in the late dreadful tempest, both by sea and land.* London: G. Sawbridge, 1704.

Derham, William. "A Letter from the Reverend Mr. William Derham, F.R.S. Containing His Observations concerning the Late Storm." *Philosophical Transactions* 24 (1704–5), 1530–34.

Dunlop, Storm. "Wind Words." *Weatherwise,* November–December 2000: 11 (letter).

Edgeworth, Richard Lovell. "Experiments upon the Resistance of the Air." *Philosophical Transactions* 73 (1783): 136–43.

Flavin, Christopher. "Harnessing the Winds." *Weatherwise,* October 1982: 211–13.

Forrester, Frank H. "Winds of the World." *Weatherwise,* October 1982: 204–10.

Hadley, George. "Concerning the Cause of the General Trade-Winds." *Philosophical Transactions* 39 (1735–36): 58–62.

Halley, Edmund. "An Historical Account of the Trade Winds, and Monsoons, . . ." *Philosophical Transactions* 16 (1686–92): 153–68.

Horn, Henry, Ran Nathan, and Sarah Kaplan. "Long-Distance Dispersal of Tree Seeds by Wind." *Ecological Research* 16 (2001): 877–85.

Jurin, James. "A Call to make Meteorological Observations." *Philosophical Transactions* 32, no. 379: 422–27.

———. *Daily Meteorological Observations,* 1 *Jan.* 1727–19 *Mar.* 1749–50. Manuscript, Royal Society, location MA. 148.

Karapiperis, Photios. "The Tower of the Winds." *Weatherwise,* June 1986: 152–54.

Keller, David. "Wind Speed Estimation Based on the Penetration of Straws and Splinters into Wood." *Weatherwise,* October 1976: 228–32.

Lind, Dr. James, Physician, at Edinburgh. "Description and Use of a Portable Wind Gage." *Philosophical Transactions* 65 (1775): 353–65.

McIlveen, J. F. R. "The Everyday Effects of Wind Drag on People." *Weather* 57 (November 2002): 410–13.

Miller, Denning. *Wind, Storm and Rain.* New York, 1952.

National Oceanic and Atmospheric Administration. *National Weather Service Observing Handbook No.* 1, *Marine Surface Weather Observations.* 1999.

Null, Jan. "Winds of the World." *Weatherwise,* May–June 2000: 36–41.

Reed, Mary. "Playing in the Wind." *Weatherwise,* May–June 2000: 49.

Vonnegut, B. "Chicken Plucking as Measure of Tornado Wind Speed." *Weatherwise,* October 1975: 217.

Watson, Lyall. *Heaven's Breath.* New York: Morrow, 1984.

Wise, A. F. E. "Effects Due to Groups of Buildings." *Philosophical Transactions of the Royal Society of London, Series A: Math and Physical Sciences* 269 (1971): 481.

Regarding John Smeaton and Alexander Dalrymple

Cook, Andrew S. "Establishing the Sea Routes to India and China: Stages in the Development of Hydrographical Knowledge." In *The Worlds of the East India Company,* edited by H. V. Bowen, Margarette Lincoln, and Nigel Rigby. Suffolk, England: Boydell Press, 2002.

Cresy, Edward. *An Encyclopaedia of Civil Engineering, Historical, Theoretical, and Practical.* London: Longman, 1856.

Dalrymple, Alexander. [Charts and Memoirs published by Dalrymple at the charge of the East India Company], 1769–1809.

———. *Practical Navigation,* ca. 1790. Printer's proof. National Library of Scotland, shelfmark Nha.M90(3).

David, Andrew C. F. "Alexander Dalrymple and the Emergence of the Admiralty Chart." In *Five Hundred Years of Nautical Science:* 1400–1900. Greenwich: National Maritime Museum, 1979.

Edwards, E. Price. *The Eddystone Lighthouses (New and Old). With an Abridgment of Smeaton's Narrative of the Building of the Old Tower.* London: Simpkin, Marshall, & Co., 1882.

Fry, Howard T. *Alexander Dalrymple and the Expansion of British Trade.* Toronto: University of Toronto Press, 1970.

Heppel, John Mortimer. "On the Relation between the Velocity and the Resistance Encountered by Bodies Moving in Fluids." *Minutes of the Proceedings of the Institution of Civil Engineers* 6 (March 17, 1847): 289–99.

Smeaton, John. *Experimental Enquiry concerning the Natural Powers of Wind and Water to Turn Mills and Other Machines Depending on a Circular Motion* . . . London, I. & J. Taylor, 1794.

———. *John Smeaton's diary of his journey to the Low Countries,* 1755. Leamingtons Spa, 1938.

Smeaton and Lighthouses. London: Parker, 1844.

Woolrich, A. P. "John Farey and the Smeaton Manuscripts." In Smith, Norman, ed. *History of Technology, Tenth Annual Volume,* 1985. London: Mansell, 1986.

Regarding the weather and the history of its understanding

Abdill, David. *A New Theory of the Weather, and Practical Views on Astronomy.* Wheeling: By the author, 1842.

Ahrens, C. Donald. *Meteorology Today.* 6th ed. Pacific Grove, CA: Brooks/Cole, 2000.

Berry, F. A. Jr., E. Bollay, and Norman R. Beers, eds. *Handbook of Meteorology.* McGraw-Hill, 1945.

Burroughs, William, Bob Crowder, Ted Robertson, Eleanor Vallier-Talbot, and Richard Whitaker. *Nature Company Guides: Weather.* San Francisco: Time/Life Books, 1996.

Cullum, John. "An Account of a remarkable Frost on the 23d of June, 1783." *Philosophical Transactions* 74: 416.

Derham, Wm., and Conestt, Tho. "An Abstract of Meteorological Diaries . . ." *Philosophical Transactions* 38 (1733–34): 101–9.

Frisinger, H. Howard. *The History of Meteorology: to 1800.* New York: Science History Publications, 1977.

Gold, E. "George Clarke Simpson." *Biographical Memoirs.* 1966. 157–75.

Haynes, B. C. *Techniques of Observing the Weather.* New York: John Wiley & Sons, 1958.

Henninger, S. K. *A Handbook of Renaissance Meteorology.* Durham, NC: Duke University Press, 1960.

Herschel, Sir John F. W., ed. *A Manual of Scientific Enquiry; prepared for the use of Officers in Her Majesty's Navy and Travellers in General.* 3rd ed. London: John Murray, 1859.

Hughes, Patrick. "FitzRoy the Forecaster: Prophet without Honor." *Weatherwise,* August 1988.

Lachla, Robert. *A Paper and Resolutions in Advocacy of the Establishment of a Uniform System of Meteorological Observation Throughout the Whole American Continent.* Cincinnati: Office of the Cincinnatus, 1859. Read before the Meteorological Section of the American Association, 30th April 1858.

List, Robert J. *Smithsonian Meteorological Tables.* 6th revised ed. Washington: Smithsonian Institution Press, 1946.

Lockhart, Gary. *The Weather Companion.* New York: John Wiley & Sons, 1988.

Maury, Matthew Fontaine. *Wind & Current Charts, Series B.* January 1851.

Middleton, W. E. Knowles. *Invention of the Meteorological Instruments.* Baltimore: Johns Hopkins University Press, 1969.

Pedgley, D. E. "Pen Portraits of Presidents: Sir George Clarke Simpson, KCB, FRS. *Weather* 50, no. 10: 347–49.

Pickering, Roger. "A Scheme of a Diary of the Weather. . . ." *Philosophical Transactions* 43 (1744–45): 1–18.

Player, E. S. "Meteorological Conditions and Sound Transmission." *Quarterly Journal of the Royal Meteorological Society* 52, no. 226 (October 1926): 351–62.

Proceedings of the International COADS Winds Workshop. Kiel, Germany, 31 May–2 June 1994.

Proceedings of the International Workshop on Digitization and Preparation of Historical Marine Data and Metadata. Toledo, Spain, 15–17 September 1997 (WMO/TD-No. 957).

Reid, William, Lt. Col., C.B., F.R.S. *The Progress of the Development of the Law of Storm, and of the Variable Winds, with the Practical Application of the Subject to Navigation.* London: John Weale, 1849.

Shaw, Napier, Sir. *Manual of Meterology.* Cambridge: The University Press, 1926, esp. vol. 1, *Meteorology in History.*

Simpson, George C. "Climatic Changes." *The Nineteenth Century and After,* January 1926: 129–41.

———. "The Development of Weather Forecasting." *The Nineteenth Century and After,* April 1927: 557–73.

Smithsonian Institution. *Meteorological Stations and Observers of the Smithsonian Institution in North America and Adjacent Islands.* Washington, DC: Smithsonian, 1869.

———. *Directions for Meteorological Observations and the Registry of Periodical Phenomena.* Washington, DC: Smithsonian, 1870.

Thomas, Charles G. "300 Years of 'Networking.'" *NOAA Magazine* May–June 1980.

Thomas, John D. "For the 'Weather-Engaged,' Manna from Heaven." *New York Times,* April 28, 2002, Arts, 10.

Tyndall, John. "On the Atmosphere as a Vehicle of Sound." *Philosophical Transactions,* 1874: 183–244.

UNESCO. "ABC of Meteorology." *UNESCO Courier* 26 (August 1973).

Weather Almanac. 1st edition. Detroit: Gale's, 1974 (and all subsequent editions).

Whipple, A. B. C. *Storm.* Virginia: Time/Life Books, 1982.

Regarding the history of the Beaufort Scale

Blackadar, Alfred. "The Beaufort Wind Scale." *Weatherwise,* October 1986: 278–80.

Capper, James. *Observations on the winds and monsoons.* London: Whittingham, 1801.

Crutcher, Harold. "Wind, Numbers, and Beaufort." *Weatherwise,* December 1975: 260–71.

Curtis, Richard. "An Attempt to Determine the Velocity Equivalents of Wind-Forces Estimated by Beaufort's Scale." *Quarterly Journal of the Royal Meteorological Society* 23 (1897): 24–61.

Cutlip, Kimberly. "Move Over, Saffir/Simpson?" *Weatherwise,* January–February 2000: 10.

Forrester, Frank H. "How Strong Is the Wind? The Origin of the Beaufort Scale." *Weatherwise,* June 1986, 147–51.

Fry, Howard T. "The Emergence of the Beaufort Scale." *Mariner's Mirror* 53, no. 4 (1967): 311–13; note, 54, no. 4, 1968.

Garbett, L. G. "Admiral Sir Francis Beaufort and the Beaufort Scales of Wind and Weather." *Quarterly Journal of the Royal Meteorological Society* 52 (April 1926): 161–72.

Havinga, A. *Windwaarnemingen in Holland in de 18th Eeuw.* Rotterdam: Van Sijn & Zonen, 1948.

"Hurricane Disaster-Potential Scale." *Weatherwise,* August 1974: 169, 186.

Joint WMO/IOC Technical Commission for Oceanography and Marine Meteorology, Subgroup on Marine Climatology, Eighth Session. Asheville, NC, 10–14 April 2000, Final Report (JCOMM Meeting Report No. 2).

Kent, Elizabeth C., and Taylor, Peter K. "Choice of a Beaufort Equivalent Scale." *Journal of Atmospheric and Oceanic Technology* 14 (April 1997): 228–42.

Kinsman, Blair. "An Exploration of the origin and persistence of the Beaufort wind force scale." Annapolis, MD, 1968.

———. "Historical Notes on the Original Beaufort Scale." *Marine Observer* 39 (1969).

———. "Who Put the Wind Speeds in Admiral Beaufort's Force Scale?" *Mariners' Weather Log* 34, no. 4 (Fall 1990): 2–8 (part 1); 35, no. 1 (Winter 1991): 12–18 (part 2).

Konvitz, Joseph. "Alexander Dalrymple's Wind Scale for Mariners." *Mariner's Mirror* 69, no. 1 (February 1983): 91–93.

Lindau, Ralf. "Rapport on Beaufort Equivalent Scales." In *Advances in the Applications of Marine Climatology—The Dynamic Part of the WMO Guide to the Applications of Marine Climatology.* WMO/TD-No. 1081, JCOMM Technical Report No. 13, 2002.

"The Log-Board." *The Nautical Magazine* 10, vol. 1 (December 1832): 537.

Petersen, P. "Zur Bestimmung der Windstarke auf See" ("To Determine the Strength of Wind on the Sea"). *Aanalen der Hydrogaphie und Maritimen Meterologie,* March 1927: 69–72.

Peterson, E. W., and L. Hasse. "Did the Beaufort Scale of the Wind Climate Change?" *Journal of Physical Oceanography,* July 1987: 1071–74.

Schlatter, Thomas. "Weather Queries." *Weatherwise,* September–October 2000: 48–49.

Scott, Robert H. "An Attempt to Establish a Relation Between the Velocity of Wind and its Force (Beaufort Scale) with some Remarks on Anemometrical Observations in general." *Quarterly Journal of the Meteorological Society,* 1874.

Shaw, W. N., and G. C. Simpson. *Official Publication No. 180, The Beaufort Scale of Wind-Force, Report of the Director of the Meteorological Office.* London: Meteorological Office, 1906.

Smith, John. *The Generall Historie of Virginia, New England, & the Summer Isles, together with The True Travels, Adventures and Observations, and A Sea Grammar.* Glasgow: J. MacLehose & Sons, 1907 (facsimile).

State of Sea Booklet. Bracknell, England: National Meteorological Library, 2001.

"The Torro Tornado Intensity Scale." *Journal of Meteorology* 8, no. 79 (May–June 1983): 151–53.

Traumuller, Dr. Friedrich. *The Mannheim Meteorological Society [Mannheimer meteorologische Gesellschaft] (*1780–1795*).* Leipzig: W. Engelmann, 1885.

United States Navy Weather Research Facility. *Meteorological Wind Scales.* Washington, DC, 1962.

Verploegh, G. *Observation and Analysis of the Surface Wind over the Ocean.* Koninklijk Nederlands Meteorologish Institut, Mededelingen en Verhandelingen No. 89, 1967.

Wallbrink, Hendrik. "Historical Meteorology revived on the Bark 'Europa,'" *Meteorologische Informatie Maritiem,* January 2002: 14–15.

Wallbrink, H., and Koek, F. B. "Historische Maritieme Windshcalen tot 1947." Monograph. Koninklijk Nederlands Meteorologisch Instituut.

Wheeler, Dennis, and Clive Wilkinson, Geography Department, University of Sunderland. "From Calm to Storm: The Origins of the Beaufort Wind Scale." Unpublished paper, 2003.

Weather 37, no. 11 (November 1982): 332 (Beaufort Scale cartoon).

World Meteorological Organization Reports on Marine Science Affairs, Report No. 3, submitted by the President of the Commission for Maritime Meteorology and the WMO Executive Committee. *The Beaufort Scale of Wind Force, Technical and operational aspects,* Geneva: WMO, 1970.

Regarding the history of science

Alder, Ken. *The Measure of All Things.* New York: Free Press, 2002.

Asimov, Isaac. *Asimov's Chronology of Science and Discovery.* New York: Harper & Row, 1989.

Brush, Stephen G. *History of Modern Science: A Guide to the Second Scientific Revolution,* 1800–1950. Ames: Iowa State University Press, 1988.

Cavendish, Margaret Lucas, Duchess of Newcastle. *Grounds of Natural Philosophy.* W. Cornwall, CT: Locust Hill Press, 1996.

Concise Columbia Encyclopedia. New York: Columbia University Press, 1983.

Deacon, Richard. *A History of the British Secret Service.* London: Panther Books, 1980.

Fitzroy, Robert. *Narrative of the surveying voyages of his Majesty's ships* Adventure *and* Beagle, *between the years 1826 and 1836, describing their examination of the southern shores of South America, and the* Beagle's *circumnavigation of the globe.* London: Colburn, 1839.

Franksen, Ole Immanual. *Babbage and Cryptography, or, The Mystery of Admiral Beaufort's Cipher.* Lecture at Babbage-Faraday Bicentennial Conference. Cambridge, July 5–7 1991.

Grun, Bernard. *The Timetables of History.* 3rd revised ed. New York: Touchstone, 1991.

Hambly, Richard. *The Invention of Clouds.* New York: Picador, 2001.

Hebrard, Jean. "Inventing the Early Modern Diary." Public lecture, October 2002, University of Michigan Institute for the Humanities.

Nye, Mary Jo. *Before Big Science.* Cambridge: Harvard University Press, 1999.

Ross, Andrew. "The Work of Nature in the Age of Electronic Emission." *Social Text* 0, no. 18: 116–128.

Rusnock, Andrea. *The Correspondence of James Jurin (1684–1750).* Amsterdam: Edition Rodopi B.V., 1996.

Society for the Diffusion of Useful Knowledge. *Penny Cyclopaedia.* London: Charles Knight & Co., 1842.

Sprat, Thomas. *History of the Royal Society.* London, 1666 (facsimile edition). St. Louis: Washington University Press, 1958.

Thomas, Keith. *Man and the Natural World: Changing Attitudes in England, 1500–1800.* Oxford University Press, 1983.

Whitehead, Alfred North. *Science and the Modern World.* New York: New American Library, 1948.

Regarding artwork and the Beaufort Scale

Ball, Nelson. *Beaufort's Scale.* Kitchener, Canada: Weed/flower, 1967.

———. *Almost Spring.* Toronto: Mercury, 1999.

"The Beaufort Scale of Wind Velocity." *Yachting Magazine,* January 1966: 110.

Canoe sailors: http://fp.ocsguk.f9.co.uk/ocsg_windscale.htm.

Collyer, Peter. *Rain, Later Good.* London: Thomas Reed, 2002.

"From Calm to Hurricane," *The Outlook: An Illustrated Weekly of Current Life,* Sept. 7, 1927, p. 11.

Heaney, Seamus. "Glanmore Sonnets" from *Field Work,* 1979.

Maddox, Jerald C. "The Beaufort Wind Scale." *Art in America,* Vol. 59, Jan.–Feb. 1971, pp. 90–91.

Paterson, Don. "The Scale of Intensity," in *God's Gift to Women.* London: Faber & Faber, 1997.

Power, Mark. *The Shipping Forecast.* London: Zelda Cheatle Press, 1998.

Sallinen, Aulis. "The Beaufort Scale," Op. 56, 1986. On *To Music: Phoenix Chamber Choir Tenth Anniversary.* Vancouver: Skylark, 1993.

Savannah weather: http://www.savannah-weather.com/weather/scales.htm.

U.S. Sailing website: www.ussailing.org/portsmouth/beaufort_spc.htm.

Wayfarer Institute of Technology website: www.angelfire.com/de/WIT.

Yvart, Jacques. *L'echelle Beaufort. D'Apres les Chroniques de Nam et Loe.* Paris: Studio SM, 1985.

Yvart, Jacques, and Claire Forgeot. *The Rising of the Wind.* La Jolla, CA: Green Tiger Press, 1984.

Primary Sources Consulted

A huge collection of Beaufort's papers, letters, and journals is in the Huntington Library and Museum in Pasadena, California. The maps that Beaufort and his department created reside in the Admiralty Archive in Suffolk, England. Beaufort's original weather diaries and other meteorological marvels are stored in the Met Office, now in Devon, England, as are the original weather reports from North Shields. The archive of the Royal Society has all the original records of that remarkable group. The Library of Congress has Beaufort's copy of Dalrymple's pamphlets and a set of maps Dalrymple gave to Beaufort, on the backs of which Beaufort has drawn maps of Montevideo. You can just ask for them in the map department and they'll bring them out. Papers from Dalrymple and others from the early days of the Hydrographic Office can be found at the Public Record Office/National Archives in London. Voting records and maps of North Shields all came from the North Tyneside Public Library in Newcastle, and the big books of births, marriages, and deaths in the United Kingdom are in the Family Records Centre, London.

Note on Accuracy

I spoke with dozens of experts and helpful people in a wide variety of fields for this book, and I consulted hundreds of sources. Eventually I asked people I trusted to read the manuscript in search of error, but there are no experts in all of this stuff, and the experts I met in each of the little areas were all very busy. All of which is to say that every mistake in this book is my own. People quoted making assertions that seem preposterous or nonsensical to people who should know can be presumed to have been misquoted, or at least quoted by a writer who incompletely understood what they said. Just the same, this is a work of nonfiction: I used no composite characters, invented no quotations, and made up no events, despite the current fashion among writers of nonfiction. All people presented as alive speaking to me or to others are alive, and actually spoke. I surely got some things wrong; I didn't make anything up.

My descriptions of the Met Office Archive and library are no longer accurate; between my visits and the writing of this book, the library moved from Bracknell, Berkshire, to Exeter, Devon, and the archive was following and will probably have moved by the time of publication.

ACKNOWLEDGMENTS

I UNDERTOOK A GREAT DEAL of the research and writing of this book while I was a Knight-Wallace Fellow at the University of Michigan. My deepest thanks go to fellowship director Charles Eisendrath, to the brilliant Wallace House staff, and above all to the 2002–2003 class of Knight-Wallace Fellows, whose spirit informs every page of this book. Also at the University of Michigan I thank Professor Perry Samson, who taught me meteorology and took the time to read this book in manuscript; Franki Hand of the Special Collections Library; Mary Pedley of the Clements Map Library; Karl Longstreth of the Map Library; Patricia S. Whitesell of the Detroit Observatory; and the many other professors and students who shared their time.

Continental Airlines graciously covered the cost of my two trips to the United Kingdom, and I thank its staff and employees, above all the very generous Gordon Bethune and Kay Jennett.

The staff of every library I consulted was extremely helpful, but I especially thank the following people and institutions: Lisa Morton, executive director of the Museum of Surveying in Lansing, Michigan; Eric Hollerton at the North Tyneside Public Library; Clara Anderson at the Royal Society; Marion James, Met Office Archive supervisor, and Steve Jebson at the Met Office library; Daniel Lewis at the Huntington Library; Adrian Webb, research manager at the data center of the Admiralty Hydrographic Office; Stephen Wildman, curator of the Ruskin Library; Tony Gardner of the Special Collections/Archives of the University Library at California State University–Northridge; and especially Andrew Cook of the British Library. Charlotte Green of the BBC offered her time, and Alisa Crawford gave an informative tour of an authentic Dutch windmill. I offer special acknowledgment to Nicholas Courtney, for both his time and his excellent biography of Beaufort.

Among people at various meteorological, scientific, and historical agencies I especially thank Scott Woodruff of the National Oceanic and Atmospheric Administration and Hendrik Wallbrink of KNMI, the Dutch meteorological organization. Virtually every member of the International Commission on the History of Meteorology helped me in one way or the other, but I send special thanks to Cornelia Luedecke; Tage Andersson; Vladimir Jankovic; and its founder, James Fleming. Dennis Wheeler and Clive Wilkinson of the University of Sunderland kindly allowed me to see a copy of their article "From Calm to Storm: The Origins of the Beaufort Wind Scale" before it was published.

Barend Visser and the very helpful Jarina den Heijer helped put me on the Bark *Europa,* and Todd Jarrell brought the *Europa* to my attention, but the entire crew and the passengers, in August 2002, made the *Europa* come to life. Sailors Jeff Werner and Jerryl Herrick helped me understand some modern connections to the old ways.

Kirsten Herold gamely wrestled into English the old Danish of Tycho Brahe, Birgit Rieck translated several articles from German, and Joris Poort translated articles from the Dutch. Musician and poet Jacques Yvart provided help and music as we discussed *The Rising of the Wind,* and artist Claire Forgeot graciously granted permission to reprint one of her lovely paintings from that book. Three other poets gave their time—Don Patterson, Paul Violi, and especially Nelson Ball, who extended permission to reprint poems from his delightful *Almost Spring* (Toronto: Mercury, 1999) and *Beaufort's Scale* (Kitchener, Canada: Weed/flower Press, 1967). At Merriam-Webster I received great assistance from Michael D. Roundy.

I thank Chris Roberts in England and John Rottet in Raleigh for their help with photographing materials in libraries.

I was treated well by a raft of friends who offered food and shelter as I traveled to research this book. I send special gratitude to the Kimball family in Seattle. Sue Nelson, Richard Hollingham, and young Matthew researched ships, crafts, and music with me in London, and Joanne Episcopo offered friendship and shelter there. Chuck Salter took the time to read and offer comments on this book in manuscript and thereby improved it. Other readers who helped try to save me from myself included Joe Miller, Michael Singer, Ron French, and Lisa Pollak. Additional catering by Leigh Menconi.

My agent Michelle Tessler nurtured and developed this book in proposal form and found it not just a home but the right home. Annik LaFarge, its editor, made a contribution to this book too large for mere thanks. Her sharp eye, keen ear, and impatience with rubbish made the difference in this project. I reserve special thanks to Emily Loose, who originally acquired this book for Crown; she was a friend to this project through all its beginnings and a friend to me for longer than that.

Above all I thank June Spence, who though she is referred to in this book as my girlfriend is now, despite hearing an awful lot about Sir Francis Beaufort, my wife, and she gets a 12 on whatever uninvented scale they use to measure that.

INDEX

About the Author

SCOTT HULER is the author of *On Being Brown: What It Means to Be a Cleveland Browns Fan* and *A Little Bit Sideways: One Week Inside a NASCAR Winston Cup Race Team*, and coauthor of *From Worst to First: Behind the Scenes of Continental's Remarkable Comeback*. He has been a staff writer for the *Philadelphia Daily News* and the *Raleigh News & Observer*. His stories, reviews, and essays have appeared in such newspapers as the *New York Times* and the *Washington Post*, and have been heard on National Public Radio and Public Radio International. He lives in Raleigh, North Carolina, with his wife, the writer June Spence, and their son.